THE JOURNEY NOT THE
ARRIVAL MATTERS

By Leonard Woolf

History and Politics
INTERNATIONAL GOVERNMENT
EMPIRE AND COMMERCE IN AFRICA
CO-OPERATION AND THE FUTURE OF INDUSTRY
SOCIALISM AND CO-OPERATION
FEAR AND POLITICS
IMPERIALISM AND CIVILIZATION
AFTER THE DELUGE VOL. I
AFTER THE DELUGE VOL. II
QUACK, QUACK!
PRINCIPIA POLITICA
BARBARIANS AT THE GATE
THE WAR FOR PEACE

Criticism
HUNTING THE HIGHBROW
ESSAYS ON LITERATURE, HISTORY AND POLITICS

Fiction
THE VILLAGE IN THE JUNGLE
STORIES OF THE EAST
THE WISE VIRGINS

Drama
THE HOTEL

Autobiography
SOWING: AN AUTOBIOGRAPHY OF THE YEARS 1880 TO 1904
GROWING: AN AUTOBIOGRAPHY OF THE YEARS 1904 TO 1911
BEGINNING AGAIN: AN AUTOBIOGRAPHY OF THE YEARS 1911 TO 1918
DOWNHILL ALL THE WAY: AN AUTOBIOGRAPHY OF THE YEARS 1919 TO 1939
THE JOURNEY NOT THE ARRIVAL MATTERS:
AN AUTOBIOGRAPHY OF THE YEARS 1939 TO 1969

A CALENDAR OF CONSOLATION: A COMFORTING
THOUGHT FOR EVERY DAY IN THE YEAR

The Journey
Not the
Arrival Matters

AN AUTOBIOGRAPHY
OF THE YEARS 1939 TO 1969

∽∽∽

Leonard Woolf

A Harvest Book
HARCOURT BRACE JOVANOVICH
NEW YORK AND LONDON

Printed in the United States of America

Library of Congress Cataloging in Publication Data
Woolf, Leonard Sidney, 1880-1969.
  The journey not the arrival matters.
  (A Harvest book ; HB 323)
  Continuation of Downhill all the way; an autobiography of
the years 1919 to 1939.
  Includes bibliographical references and index.
  1. Woolf, Leonard Sidney, 1880-1969. I. Title.
JA94.W6A29 1975 320′.092′4 [B] 75-9822
ISBN 0-15-646523-X

First Harvest edition 1975

A B C D E F G H I J

CONTENTS

ILLUSTRATIONS

Chapter One

VIRGINIA'S DEATH

THE second of the great world wars through which I have lived began on September 3, 1939. Twenty-five years before, the great war of 1914 on a summer day in August had come upon us, upon our generation and indeed upon all the generations of Europe, historically and psychologically, a bolt from the blue. It was as if one had been violently hit on the head, and dimly realized that one was involved in a dreamlike catastrophe. For a hundred years a kind of civilization seemed to have been spreading over and out of Europe so that an Armageddon had become an anachronistic impossibility or at least improbability. There had been wars and we still prayed automatically on Sundays to a very anachronistic God to deliver us from "battle" as from murder and sudden death, from the "crafts and assaults of the devil" and from "fornication, and all other deadly sins"; but the wars were local or parochial wars and millions of people had lived and died without hearing the drums and tramplings of any conquest, or had had the remotest chance of standing "on the perilous edge of battle".

The psychology of September 1939 was terribly different from that of August 1914. People of my generation knew now exactly what war is—its positive horrors of death and destruction, wounds and pain and bereavement and brutality, but also its negative emptiness and desolation of personal and cosmic boredom, the feeling that one is endlessly waiting in a dirty, grey railway station waiting-room, a cosmic railway station waiting-room, with nothing to do but wait endlessly for the next catastrophe. We knew that war and civilization in the modern world are incompatible, that the war of 1914 had destroyed the hope that human beings were becoming civilized—a hope not unreasonable at the beginning of the twentieth century. The Europe of 1933 was infinitely more barbarous and degraded than that of 1914 or 1919. In Russia for more than a decade there had ruled with absolute power a government, a political party, and a dictator who, on the basis of a superhuman doctrinal imbecility, had murdered millions of their fellow-Russians because they were peasants who were not quite so poor as the poorest peasants; the communists, being communists, were continually torturing and murdering their fellow-communists on such grounds as that they were either right deviationists or left deviationists. In Italy there was established a government and dictator who, with a political doctrine purporting to be the exact opposite of Russian communism, produced, much less efficiently, exactly the same results of savage stupidity. In Germany the same phenomena had appeared as in Russia and Italy, but the barbarism of Hitler and the Nazis showed itself, in the years from 1933 to 1939, to be much nastier, more menacing, more insane than even the barbarism of Stalin and the communists.

In many ways, therefore, the last years of peace before war broke upon us in 1939 were the most horrible period of my life. After 1933 as one crisis followed upon another, engineered by Adolf Hitler, one gradually realized that power to determine history and the fate of Europe and all

Europeans had slipped into the hands of a sadistic mad-
man. When one listened on the air to the foaming hysteria
of a speech by the Führer at some rally, whipping up the
savage hysteria of thousands of his Nazi supporters, one
felt that Germany and the Germans were now infected
with his insanity. As the years went by, it became clear
that those in power in Britain and France would offer no
real resistance to Hitler. Life became like one of those
terrible nightmares in which one tries to flee from some
malignant, nameless and formless horror, and one's legs
refuse to work, so that one waits helpless and frozen with
fear for inevitable annihilation. After the Nazi invasion of
Austria one waited helpless for the inevitable war.

It was this feeling of hopelessness and helplessness, the
foreknowledge of catastrophe with the forces of history
completely out of control, which made the road downhill
to war and the outbreak of war so different in 1939 from
what they had been twenty-five years before. A few facts
connected with this foreknowledge and despair are worth
recollecting. In the year before the outbreak of the war I
was asked by Victor Gollancz, Harold Laski, and John
Strachey to write a book for the Left Book Club. I wrote a
book to which I gave the title *Barbarians at the Gate*. I
began by quoting "words written about twenty-five
centuries ago" by Jeremiah, "the father of communal
lamentation", lamenting the destruction of a civilization
by barbarians, who have burnt incense to strange gods,
filled Jerusalem with the blood of innocents, and burnt
their sons with fire for burnt offerings unto Baal—"therefore, behold, the days come that this place shall no more
be called Tophet, nor The valley of the son of Hinnom,
but The valley of slaughter". I went on to point out

(nearly thirty years ago) the difference between 1938 and 1914, for "when you opened your newspaper in those days, you did not read of the wholesale torture, persecution, expropriation, imprisonment or liquidation of tens of thousands or hundreds of thousands of persons, classified or labelled for destruction as social-democrats, communists, Jews, Lutheran pastors, Roman Catholics, capitalists, or kulaks". I insisted that the ultimate threat to civilization was not so much in this barbarism of the barbarians as in the disunity among the civilized, and I made the correct and sad prophecy:

"It is practically certain that economics, a war, or both will destroy the Fascist dictators and their regimes. But that does not mean that civilization will automatically triumph over barbarism."

I have one amused recollection connected with that book. When I sent in the MS, I received a troubled letter from Victor; the three editors liked the book very much, he said, but they were worried about my criticism of the Soviet Government and communism, would I consider toning it down? I replied that I was willing to consider any precise and particular criticisms or suggestions for alterations, but I was not prepared to modify my views on the grounds of expediency. In the end it was decided that we should all four meet and discuss the MS in detail face to face. I met the editors in Victor's office after dinner on July 24, 1939. They were much upset by my criticism of the Russian communists and their government and pressed me to modify it. The modifications which they asked for would, I felt, be dishonest, from my point of view, for they would obscure what, in my opinion, was the truth about authoritarianism in the Russia of Stalin. The barbarians were already within the gates both in Moscow and Berlin; to conceal or gloss over the truth in the communist half of Europe would make disingenuous nonsense of my book. I refused to budge, and the discussion went on for two or three hours, becoming more and more difficult, as warmth increased on their side of the room and frigidity on mine. The book was published unaltered and unmodified, and I can console or even congratulate myself that today, if they were alive—alas, all three are dead— my three editors would agree with everything which I wrote.

That evening when I got up to go, I felt that I had not ingratiated myself with my three friends, all of whom I liked very much, both in private and public life. There was a slight cloud, slight tension in the room, but I am glad to remember that, before I went out, an absurd little incident entirely dispersed them. On the wall opposite to where I had been sitting was a picture which through the long, rather boring and exasperating, argumentative discussion I had frequently looked at with pleasure and relief. In gratitude to the painter, when I said good-night to Victor, I asked him who had painted it and added that I had got a great deal of pleasure by looking at it. I could not have said anything to give Victor more pleasure or more effectively relieve the tension, for the painter was his wife. I left the room, not under a cloud, but in a glow of good-will and friendship.

To return to the psychology of the years before the war, one only gradually became aware of the savage barbarism of the Nazis in Germany and of the inevitability of war, but I still remember moments of horrified enlightenment.

When a Jew shot a German diplomatist in Paris, the Government instigated an indiscriminate pogrom against Jews throughout Germany. Jews were hunted down, beaten up, and humiliated everywhere publicly in the streets of towns. I saw a photograph of a Jew being dragged by storm troopers out of a shop in one of the main streets in Berlin; the fly buttons of the man's trousers had been torn open to show that he was circumcised and therefore a Jew. On the man's face was the horrible look of blank suffering and despair which from the beginning of human history men have seen under the crown of thorns on the faces of their persecuted and humiliated victims. In this photograph what was even more horrible was the look on the faces of respectable men and women, standing on the pavement, laughing at the victim.*

As I recorded in *Downhill All the Way*,† when I drove through Germany in 1935, "we did not enjoy this; there was something sinister and menacing in the Germany of 1935. There is a crude and savage silliness in the German tradition which, as one drove through the sunny Bavarian countryside, one felt beneath the surface and saw, above it, in the gigantic notices outside the villages informing us that Jews were not wanted." In 1935, however, Hitler had been in power only two years, and one felt only vaguely "this crude and savage silliness" beneath the sur-

* I feel something peculiarly terrible in the static horror of that photograph, the frozen record of the victim's despair and the spectators' enjoyment. Even more horrible and haunting is a photograph, published after the war, of a long line of Jews, men, women and children, being driven naked down a path into a gas chamber. Here again one sees visually before one the barbarism of the human race in the middle of the twentieth century.

† Page 192.

face. There was in fact something much more savage and sinister beneath the surface, and in the next few years one occasionally caught a glimpse of it. For instance, just before the war Adrian Stephen learnt that a German friend of his was in grave danger from the Nazis, and, with influential support behind him, he went to Berlin to try to get his friend out of Germany. He conducted some very strange and complicated negotiations in the course of which he saw something of what the Nazis were doing and meant to do, and also of the desperate plight of their victims. The account of his experiences still further opened one's eyes—one looked into the abyss. As a result of that vision of the German brutality, when war had actually come and one had to face the possibility, if not probability, of invasion, Adrian told us that he would commit suicide rather than fall into German hands, and that he had provided himself with means of doing so; he offered to Virginia and me, who would certainly have been among the proscribed, a portion of this protective poison. I gather from Harold Nicolson's memoirs that he and Vita provided themselves with a similar "bare bodkin", so that they might make their quietus in order to avoid the fate which would be theirs if they fell into German hands. Here again is terrible evidence of the difference in savagery between the Europe of 1939 and 1914. For here in 1939 were five ordinary intelligent people in England, coolly and prudently supplying themselves with means for committing suicide in order to avoid the tortures which almost certainly awaited them if the Germans ever got hold of them. It is inconceivable that anyone in England in 1914 would have dreamt of committing suicide if the Kaiser's armies had invaded England.

In writing an autobiography covering the years 1939 to 1945 one should, I think, try somehow or other objectively to face the facts about the horrible savagery of Hitler and the Germans. There was something insane in Hitler's genocide; in his writings, recorded conversation, and acts; in the conception and the execution of his colossal plan for killing in cold blood millions of human beings merely because they belonged to a race or religion which he did not like—Jews, Poles, or gipsies. But this sadistic nightmare of an insane megalomaniac was and could be executed only by hundreds, by thousands of ordinary sane Germans. They killed in various ways, but mainly by driving into lethal gas chambers six or seven million human beings with the greatest efficiency and the most appalling cruelty. The doctors who performed their disgusting experiments and operations on their victims, the commandants and guards who year after year starved and tortured millions of their fellow-citizens in the German concentration camps, seemed to be infected with Hitler's sadistic insanity. One hears occasionally quite casually of facts which show how widespread among the Germans was this inhuman cruelty. A Dutchman, a manual worker, told me that, when the Germans occupied Holland, he had to work for them on an airfield. On a railway line near by they loaded Jews into cattle trucks to be taken off to Germany, where they were destroyed. One day he saw a small child, frightened and crying, pull away from his mother so that she could not get up into the truck. A German guard caught hold of the child by one leg and flung him, as if he were a sack of corn, up into the air and over the side so that he fell onto the floor of the truck. The Dutchman told me that he could never

16

forget the sight; it haunted him; it made him hate all Germans.

"It made him hate all Germans"—the sentence haunts me, just as the face of the well-dressed woman in the photograph laughing at the tortured face of the Jew with his fly buttons torn open and the bewildered faces of the naked women and children in the other photograph being driven down the narrow valley by the German uniformed guards into the death chamber haunt me. The callous cruelty, the pitilessness, the dreadful senselessness of those persecutors and murderers, and the hatred of all Germans which they generated in the Dutchman are the stigmata of the world in which I have lived since 1914. I feel the hatred welling up in myself, and yet I hate the hatred, knowing it to be neither rational nor objective. There is an old well-worn tag which says that one cannot condemn a nation, and there is some truth in it. Yet the scale of German cruelty and barbarism under Hitler in the years from 1933 to 1945 is so colossal that it seems to be different in quality or kind from the barbarism of other European peoples.

These horrible events and their effect upon personal and communal psychology in the world in which I have had to spend my life seem to me of profound importance; to understand them is also profoundly important. To understand them, at least to some extent, one must, I think, consider the nature and history of cruelty. Montaigne in one of his essays writes:

Amongst all other vices there is none I hate more than cruelty, both by nature and judgement, as the extremest of all vices. But it is with such an yearning and fainthartednesse, that if I see but a chickins neck pulled off, or a pigge stickt, I cannot chuse but grieve, and I cannot well endure a seelie, dew bedabled hare to groane, when she is seized upon by houndes.

I agree with Montaigne, there is nothing more horrible in human beings than human cruelty. But it is not just a question of liking or disliking or tolerating a vice or a virtue. I am writing these words in September 1967; it is four hundred years ago that Montaigne wrote the sentence quoted above—he may well have been sitting in the tower on the wall of his château in Montaigne on a September morning of 1567 when he wrote it. He was, I think, the first person in the world to express this intense, personal horror of cruelty. He was, too, the first completely modern man; he was pre-eminently a man of the Renaissance, that movement in the minds of men and therefore in history which created a new civilization, modern civilization which began in the Renaissance of the fourteenth century and was destroyed in 1914. An integral part in that new civilization was the revolution in man's attitude to man. Before the Renaissance in all previous civilizations the individuality of the individual human being was only dimly realized and counted for little or nothing in the ethics and organization of society; men, women, and children were not individuals, were in no sense "I's", they were anonymous, impersonal members of classes or castes. In the middle of the fourteenth century this medieval attitude towards human beings, which was the basis of medieval society, began to give way to an uneasy awareness of the individuality of the individual. Montaigne was the first completely modern man in his intense awareness

18

of and passionate interest in the individuality of himself and of all other human beings.

The combination in Montaigne of intense hatred of cruelty and intense awareness of individuality is not fortuitous. There is no place for pity or humanity in a society in which human beings are not regarded as individual human beings, but as impersonal classified pegs in a rigidly organized society. It is only if you feel that every he or she has an "I" like your own "I", only if everyone is to you an individual, that you can feel as Montaigne did about cruelty. It is the acute consciousness of my own individuality which makes me realize that I am I, and what pain, persecution, death means for this "I". For me "death is the enemy", the ultimate enemy, for it is death which will destroy, wipe out, annihilate me, my individuality, my "I". What is so difficult to understand and feel is that all other human beings, that even the chicken, the pig, and the dew bedabled hare, each and all have a precisely similar "I" with the same feelings of personal pleasure and pain, the same fearful consciousness of death that destroyer of this unique "I". In the civilization which developed from the Renaissance the ultimate communal ideal was defined in the famous liberty, equality, fraternity of the French Revolution. But those words only translate into social and political terms the consciousness of universal individuality and the right of everyone to be treated as an individual, a free fellow-human being. The development of a civilization—its beliefs, ideals, and institutions—is a long and oscillating process. In the years between the life of Montaigne and the 1914 war there was a continual ebb and flow in the struggle for the emergence of the individual, for the right of everyone to be treated equally as an "I"—even the baited bull and the hunted hare. By 1900 a civilized society, based upon individuality and liberty, equality and fraternity, had established itself with some firmness in only a few places, but taking the world as a whole nearly everywhere the movement of events and in men's minds seemed, despite oscillations, to be in the direction of Montaigne and Erasmus, Voltaire and Tom Paine.

Born in 1880 and bred in the bourgeois Kensington house and the "liberal" atmosphere which derived from my father's sensibility, his extreme physical and moral fastidiousness, I very early came to understand and feel in my bones and recognize in events the various manifestations of this civilization and of the counter-revolution still fighting bitterly against it. The spectrum of this civilization stretched from political and social democracy at one end to humanitarianism and the Society for the Prevention of Cruelty to Animals at the other. What it means, and what it meant to me personally, and how its various manifestations are deeply rooted in our attitude to individuality and individuals can best be shown by a few examples of great oecumenical events and personal egocentric experiences which profoundly influenced or affected me.

First what some people may think a trivial and sentimental recollection, but an incident which to me is of great importance carrying me back to Montaigne and the essential nature of civilization. It happened to me years ago when I was a boy and had not yet read the passage quoted above from Montaigne. My bitch had five puppies and it was decided that she should be left with two to bring up and so it was for me to destroy three. In such circumstances it was an age-old custom to drown the day-old puppies in a pail of water. This I proceeded to do. Looked at casually, day-old puppies are little, blind, squirming, undifferentiated objects or things. I put one of them in the bucket of water, and instantly an extraordinary, a terrible thing happened. This blind, amorphous thing began to fight desperately for its life, struggling, beating the water with its paws. I suddenly saw that it was an individual, that like me it was an "I", that in its bucket of water it was experiencing what I would experience and fighting death, as I would fight death if I were drowning in the multitudinous seas. It was I felt and feel a horrible, an uncivilized thing to drown that "I" in a bucket of water.

From the blind puppy struggling in the bucket of water to thousands of men, women, and children struggling for life in the mountains of Asia Minor. In 1894 one of those savage and senseless internecine massacres, epidemic among human beings, broke out in the Ottoman Empire. Turks and Kurds, encouraged by the Ottoman Government, began a systematic looting and destruction of Armenian villages and the slaughter of the inhabitants. The motives were religious, racial, and economic—which means that they were senseless, uncivilized, and inhuman. To kill a man and his wife, to rape his daughter and then kill her, because they pray in a church instead of a mosque, talk Armenian instead of Turkish, and are slightly (or thought to be slightly) more prosperous than you are, is senseless and barbarous, and the motives given above are labels which conceal a deeper and darker part of the human mind. The man who massacres can only do this if he regards his victims not as individuals like himself but as non-human pawns or anonymous ciphers in the fantasy or nightmare world of friends and foes, good men and evil men, in which he thinks he lives and which he therefore creates—or, of course, if he is just a plain common or garden sadist.

The campaign against Armenian massacres, like the antislavery movement of the early nineteenth century, was an example of a sudden mass movement against barbarism. It was led in England by Gladstone. That strange, Jesuitical, passionate, human man was eighty-five years old and had retired from politics, but he came out of his retirement and in a series of great public speeches denounced the massacres and their instigators as a disgrace to civilization and called upon the Disraeli Government to intervene and stop them. I was fourteen years old at the time and it was my first profound political experience. Gladstone's campaign had a tremendous effect upon all kinds of different people, among whom was Mrs Cole, the headmistress of the school to which my sisters went and which in its kindergarten first introduced me to school life and the stirrings of sex.* Mrs Cole was either the apotheosis or the caricature of the Victorian female teacher, depending upon the angle from which you observed her. She dominated the school and everyone in it, including her husband, who looked like the Prince Consort. Appropriately and, I think, deliberately she looked slightly like Queen Victoria. She was a short dumpy woman with thick shiny black hair, parted in the middle, and drawn back tightly and smoothly to a bun at the back. She was always dressed in black silk but somewhere or other on her person was a flounce or flower of bright pink.

* See *Sowing*, p. 53.

On her head, even when she was in the house, there was always a black bonnet with two long broad black ribbons attached to it. A woman of immense energy, mental and physical, she was perpetually whizzing about the house up and down stairs, in and out of classrooms, with the two black ribbons billowing out behind her. She addressed everyone, including whole classes and her flabby husband, in tones and vocabulary of cooing endearment—but there was an iron will under the velvety voice.

Mrs Cole became obsessed with the horrors and barbarism of the Armenian massacres. She descended like a whirlwind of black silk ribbons on her friends and acquaintances, imploring them to call upon the British Government to stop the massacres, beseeching them to give money to the Armenian Fund and woollen socks, stockings, and mittens to the starving and freezing survivors. The terrible stories and Mrs Cole's passionate indignation had a great effect upon me: for the first time I had, I think, a vague feeling or dim understanding of the difference between civilization and barbarism. I could almost see the helpless Armenians being bayoneted by the Turkish soldiers and the women and children fleeing and floundering through the snowdrifts. And I had a shadowy feeling—as I had about the puppy in the bucket—that each of these victims was a person, like me an "I".

Thirdly, to the puppy drowning in the bucket and to the massacred Armenians I must add the tragic figure of Captain Dreyfus. I have already in *Sowing*** said something about the Dreyfus case as a key event in the history of our times, a symbol of the eternal struggle between barbarism and civilization. Here I am concerned with the revelatory

*Pages 152 and 161-2.

effect upon me which it shared with the puppy and the Armenians. If you look up Dreyfus in the current edition of the famous French dictionary *Petit Larousse*, published over seventy years after the obscure Captain Dreyfus was convicted by a court-martial of espionage and condemned to life imprisonment, you will find the following entry:

> Dreyfus (Alfred), officier français, né à Mulhouse (1859-1935), Israélite, accusé et condamné à tort pour espionnage (1894), il fut gracié (1899) et réhabilité (1906), après une violente campagne de révision (1897-1899) dénaturée par les passions politiques et religieuses. Ses adversaires étaient groupés dans la ligue de la Patrie française, ses partisans dans celle des Droits de l'homme. L'affaire avait divisé la France en deux camps.

These few lines give with admirable exactitude the bare bones of the Affaire: they put in the forefront that Dreyfus was Israélite—a Jew; an innocent man, he was accused and convicted of espionage; five years later he was given a free pardon, and, seven years later, after a terrific campaign for revision, he was retried, found to be innocent, and reinstated. The case divided France into two camps, for the battle against and for Dreyfus was fought with intense bitterness, between, on the one side, the army, Church, and conservatives and, on the other, the liberals and radicals.

It took a considerable time before people living outside France became aware of the Dreyfus case and its importance, and so it was several years after the Armenian massacres that I had my second political revelation from Paris. For some time after the conviction in 1894 one accepted the fact that a French officer had been convicted and sentenced for espionage, but by 1899 and the second court-martial one had become convinced of Dreyfus's innocence and one saw what his case involved for the future of France and of civilization. It seemed in those days a terrible thing that the vast power of the State, the army, the Roman Catholic Church, and the press, concentrated in the hands of Cabinet Ministers, generals, cardinals, bishops, and editors, should deliberately be used by them to conceal and pervert the truth in order to ensure the conviction and life imprisonment of a man for a crime which they knew he had not committed. Occasionally in history some solitary figure in some tragic scene is transfigured and becomes a symbol of innocence or sin, of compassion or cruelty, of victory or defeat, of civilization or barbarism. The Israelites three or four thousand years ago, in their passionate preoccupation with sin, invented such a symbolic scene in which the priest

> shall lay both his hands on the head of the live goat and confess over him all the iniquities of the children of Israel, and all the transgressions in all their sins, putting them on the head of the goat, and shall send him away by the hand of a fit man into the wilderness; and the goat shall bear upon him all their iniquities unto a land not inhabited: and he shall let go the goat in the wilderness.

The most famous of all such symbolic scenes took place nearly two thousand years ago in Jerusalem and is itself connected with the obsession of the Jews with sin and of the Christians and their churches who have absorbed and developed this obsession. The man accused before Pilate and condemned to be crucified between the two thieves is transfigured as the son of God, the symbol of innocence,

25

salvation, civilization, while the barbarians hissing and shouting, cry: "Crucify him! His blood be on us and on our children." And, as I pointed out in *Sowing**, the scene of the formal degradation—a kind of crucifixion—of Dreyfus acquired the same kind of symbolic import— Dreyfus is brought into the huge square formed by detachments of all the regiments; the general says to him: "Alfred Dreyfus, you are unworthy to bear arms. In the name of the French People we here degrade you". Dreyfus raises his arms and cries: "Soldiers, I am innocent! It is an innocent man who is being degraded, Vive la France! Vive l'Armée!"; a sergeant tears off from Dreyfus's uniform the insignia of his rank and breaks his sword, while the crowd hiss and shout: "Kill him! Kill him!"

Because of all this Dreyfus had an even greater effect upon me than the drowning puppy and the massacred Armenians. The case was symbolic in two different ways, and to watch the interminable, fluctuating struggle was doubly agonizing. There was first involved the impersonal, general principle of justice. People differ a great deal in their feeling about justice in the abstract. To many people it seems to mean little or nothing; others—and I am one of them—agree with the man, whoever he was, who said: *"Fiat justitia, ruat coelum"*. My father, as I recalled in *Sowing*,† thought that a perfect rule of conduct for a man's life had been laid down by the prophet Micah in the words "do justly and love mercy". I think that I too have always felt intensely about this. I get a keen kind of aesthetic pleasure in a complicated case in which perfect justice is done; on the other hand, injustice of any kind or to any person is extraordinarily disturbing and painful.

*Pages 161-2.   †Page 26.

An unjust law or a miscarriage of justice hurts and jars me like a false quantity or a discord in the wrong place, or a bad poem, picture, or sonata, or the stupidity of the overclever, or the perversion of truth. In all these cases the pain is impersonal, though it is no less acute for being that. But in cases like that of Dreyfus there is a second element which fills one with horror and despair. It is not merely that the impersonal principle of right or wrong, justice or injustice, is involved; Socrates condemned to death in Athens, Christ crucified in Jerusalem, Calas* condemned and tortured to death by the Church in France, Dreyfus condemned in Rennes and tortured in Devil's Island—in all these cases a person, an individual, faces us with the terrible, accusing, symbolic cry, addressed to God or to man, to society, to civilization: My God! My God! why hast thou forsaken me? Dreyfus was not merely an anonymous unit among anonymous units, soldiers, officers, captains, Jews, like me and like the puppy—and Socrates, Christ and Calas—he was an "I", an individual, and it was as this "I" that a civilized society

* The case of Calas, falsely accused by the Roman Catholics of having murdered his son in order to prevent him renouncing Protestantism, was in every respect the Dreyfus case of the eighteenth century. Voltaire, in his passionate plea for the right of the individual to justice, did exactly what Zola did 140 years later, and the dead Calas was "rehabilitated" in 1765 just as the living Dreyfus was "rehabilitated" in 1906. Voltaire made the Calas judicial murder a test case of civilization. To quote Theodore Besterman in *Voltaire Essays:* "His cry of indignation echoed round the world finally triumphed ... For the first time and for good since the day on which Voltaire opened his arms to the Calas family ... social justice has been on the defensive: and it will remain on the defensive so long as men retain Voltaire's gift of indignation" (Mr Besterman's last sentence is, I think, over-optimistic).

regarded him, his innocence or his guilt, his punishment and sufferings: not to do so is the negation of civilization.

I am insisting on this because it is essential to an understanding of the difference between the political and historical climate of 1939 and that of 1914; it also explains why people of my generation regarded with despair the world which Stalin, Mussolini, and Hitler had made, why so many people watched the war inevitably coming and entered it with a strange mixture of misery, calmness, and resignation. We knew that in Russia, Italy, and Germany there were hundreds of Calases, thousands of Dreyfuses. The world had reverted to regarding human beings not as individuals but as pawns or pegs or puppets in the nasty process of silencing their own fears or satisfying their own hates. It was impossible even for that most savage of all animals, man, to torture and kill on a large scale peasants, fellow-socialists, capitalists, Jews, gipsies, Poles, etc. if they were regarded as individuals; they had to be regarded as members of an evil and malignant class—peasants, deviating socialists, capitalists, Jews, gipsies, Poles. The world was reverting or had reverted to barbarism.

Personally I had felt this deeply and bitterly all through the last two years before war broke out. There was the horrible ambivalence towards Chamberlain's shameful betrayal of Czechoslovakia. Chamberlain always seemed to me the most coldly incompetent, most ununderstanding, unsympathetic of the British statesmen who have mismanaged affairs during my lifetime. But when one stands on the very brink of war and suddenly, when one has practically abandoned hope, there is a shift in the kaleidoscope of events to peace instead of war, one cannot but feel an immense relief, release, and reprieve, even though at the same time one feels that the steps which have led to the avoidance of war ought not to have been taken, being shameful and morally and politically wrong. I suffered from this ambivalence all through the Munich crisis, for though the relief was extraordinary, I was convinced that by abandoning Czechoslovakia to Hitler we were only postponing war and that when it came we should have to fight it under conditions far more unfavourable to us than if we had Czechoslovakia and Russia as allies.

When the Polish crisis started, I felt that the end was coming. We went down to Monks House for the summer on July 26, but on August 17, just seventeen days before we were actually at war, we had to move from Tavistock Square to the house which we had taken in Mecklenburgh Square; so we had to be driving backwards and forwards between Rodmell and London. The air of doom and calm resignation both inside one and outside one is what I chiefly remember of those days. The appearance of sandbags, the men digging trenches, the man on the removal van taking our furniture from Tavistock Square to Mecklenburgh Square and as an ex-soldier receiving his call-up notice (I shan't be here tomorrow, Sir)—all with this quiet, dull, depressed, resigned sense of doom.

I suppose that, from the beginning of human history, men and women, the nameless individuals, have always faced the great crises and disasters, the senseless and inexorable results of communal savagery and stupidity, with the calm, grim, fatalistic resignation of the furniture removal man and all of us in Rodmell and London in August and September 1939. It is a kind of sad consolation to think that it must have been almost exactly like this in Athens in August and September 480 B.C. Ten years before, the horrors of war and invasion had swept down upon the Athenians with the Persian armies within a few miles of Athens. All the men of military age were called up, ten thousand infantry—"I shan't be here tomorrow, Sir"—and the Persian army was defeated and the war ended with the battle of Marathon. It answered to our war of 1914-1918. Then the great army and fleet of Xerxes began again in 480 B.C. just as Hitler began again in 1939. All the men of military age were once more called up—"I shan't be here tomorrow, Sir". The Persian armies swept down after the fall of Thermopylae (as France fell in 1940) and the whole population of Athens was evacuated to Salamis and other islands, just as the population of the East End of London was evacuated to Rodmell and other villages in 1939. Every Athenian who had been above the age of fourteen or fifteen in 490 must have remembered vividly in 480 the horrors and terrors of the first war and invasion. The young men once again drafted into the Greek army or navy, the old men and women ferried across from the Piraeus to Salamis were each and all of them individuals, as we were in London, waiting for the blitz. And I am sure each of them was feeling the same depressed resignation which Virginia and I felt driving up to London or watching a gang of Irish labourers digging, with incredible slowness and indolence, a bomb shelter in Mecklenburgh Square.

We were in Rodmell on Sunday, September 3, when war was declared. The people evacuated from Bermondsey arrived and we helped to settle them into the cottages hurriedly prepared for them. They were typical Londoners and nearly all of them were horrified by our cottages and outraged and enraged at being asked to live in them. Most of them within a week or two had packed up and left us; they preferred to face the risk of Hitler's bombs to life in a Sussex village. The strange first air raid of the war—it was, of course, a false alarm—came to Rodmell on a lovely autumnal or late summer day. It came, I think, just after or before breakfast and I walked out onto the lawn which looks over the water-meadows to Lewes and the downs. It was absolutely still; soft, bright sunshine with wisps of mist still lying on the water-meadows. There are few more beautiful places in England than the valley of the Sussex Ouse between Lewes and Newhaven, the great sweep of water-meadows surrounded by the gentle, rounded downs. On a windless summer morning the soft sparkle of the sun on the meadows and downs, and every now and again the narrow ribbons of white mist lying upon them, give one an extraordinary sense of ageless quiet, King Arthur's "Island valley of Avilion"—though one could never describe our water-meadows as a place

Where falls not hail, or rain, or any snow
Nor ever wind blows loudly.

It is curious that this Ouse valley should be so visually connected in my mind with peacefulness and beauty while I listened to the first air-raid sirens of the 1939 war, for, during the next six years, as soon as the phoney war ended and the real war began, it was over the peaceful water-meadows and above our heads over Rodmell village that again and again I watched the many strange phases of the war in the air being fought. Perhaps this is the best place to say something about how the air fighting actually affected us in the Sussex countryside.

31

The first sight of German planes which we saw was very odd. The real air war began for us in August 1940. On Sunday, August 18, Virginia and I had just sat down to eat our lunch when there was a tremendous roar and we were just in time to see two planes fly a few feet above the church spire, over the garden, and over our roof, and looking up as they passed above the window we saw the swastika on them. They fired and hit a cottage in the village and fired another shot into a house in Northease. This first experience of active warfare surprised me in one respect. I had always thought that I should be frightened under fire. The German planes just above my head, I was glad to find, left me perfectly calm and cold, the whole incident seeming to be completely unreal, and in fact in all the many "incidents" of the kind which took place in subsequent years I never myself felt or saw anyone else feeling fear. Though between 1940 and 1945 I must have seen hundreds of German planes and many of them dropping bombs or fighting British planes, except in this incident I never saw or had real evidence of a German plane firing bullets at people or buildings on the ground, but very early in the war I had an experience with regard to the alleged machine-gunning by a German airman which taught me never to believe stories of "incidents", even by eye-witnesses. It is worth recording.

In those distant days newspapers were delivered in villages like Rodmell by men or girls riding bicycles. There were two delivering in Rodmell whom I knew well. One was a man whom I will call Tom and the other was a girl of about 17 whom I will call Mary. One day I realized that it was several weeks since I had seen Mary and her bicycle, and a day or two later I met Tom in the village street. "What has happened to Mary, Tom?" I said; "I haven't seen her for a long time." "Haven't you heard?" said Tom. "She was machine-gunned by a German plane, and I saw it happen. I was riding on the Ringmer road and Mary was riding about two hundred yards ahead of me. Suddenly a German plane swooped down, flew just above our heads along the line of the road, and fired a burst at Mary. She was badly wounded and is still in the hospital." Here, I thought, at any rate is an authentic case of a senseless and brutal German "atrocity". About three months later I met Mary walking along a street in Lewes. I congratulated her on her recovery, but as soon as she began to talk about her experience, I found that the story told by Tom was entirely untrue. There had been no German plane and no shooting and the incident had taken place not in Ringmer but in the Lewes street where we were talking. Mary had been walking up the street during an air raid and when she was immediately opposite the Co-op store a bomb fell on a house not many yards away. The blast shattered the window of the shop and Mary was badly cut by the flying glass. I find it extremely difficult to understand Tom's behaviour in inventing, as he must have done, a completely false story. I knew him quite well and often talked to him; once when he was ill and in hospital, I went to see him and we talked about all kinds of things for half an hour or more. He never gave the slightest sign of "talking tall" or of lying. I am inclined to think that by the time he told me that he had seen Mary shot on the Ringmer road, he really believed that he had. Yet he must at the same time have known that she was in Lewes hospital having been badly cut by flying glass from the shattered Co-op window. "There's fevers of the mind as well as body", as Mrs Gamp remarked, and in wartime they show themselves in the wild way in which people invent stories of what they have heard or seen.

To return to bombing and to what did happen, some time in 1940 there were German planes over Rodmell one day and Virginia and I were standing in the garden when we heard the swishing of bombs through the air overhead and then the dull thuds of explosions towards the River Ouse. The bombs were aimed at and missed the cement works, but one or two of them hit and breached the river bank. There happened to be quite a high tide at the time and the river poured through the gap and flooded the fields. Then some days later there was an abnormally high tide with a strong wind, and a great stretch of river bank gave way. The whole Ouse valley was flooded and a great lake of water now stretched from the bottom of my garden to Lewes on the north and almost to Newhaven on the south. We had reverted to the conditions of the early nineteenth century before the river banks were built up. In those days whenever there was heavy rain and a high tide the whole valley was under water.

Early in the war I joined the fire service. My duties consisted of taking my turn patrolling the village at night and also helping to man a pump. The pump was a beautifully primeval machine which was mounted on a small truck. Four firemen, yoked like horses, pulled the truck to the (presumably) blazing building on which we hoped to direct a stream of water through long hoses. We used in practice to rush this strange contraption up and down the village street and play the water on some house or cottage. We were only once called out in earnest; about ten o'clock one night in March 1941, during a raid, a German bomber dropped a shower of incendiary bombs which just missed the village and fell all over the fields on the south of it. One of them hit a haystack a few hundred yards beyond the last house on the road to Newhaven. We rushed our pump to the spot, but the heat of the burning stack was so terrific that we could not get anywhere near it with our length of hose. So we rang up the professional fire engine in Lewes; when it arrived it was as impotent as we were, for by the time they arrived two other stacks had caught and even the professionals could do nothing. It was a curious sight typical of the way in which in the second world war people in Sussex faced battle, murder, and sudden death. Though there was an air raid in progress, half the village turned out and stood on the main road watching the burning ricks. Scattered over the fields there were numbers of little "flames upturning" where the bombs were burning themselves out. We—twenty or thirty men, women, and children—the firemen, and the fire engines stood on the road in the tremendous glare of the fire. Suddenly came the drone of a plane and a German bomber flew low over our heads. I think everyone expected a bomb or two, but nobody moved. The plane plugged away and some minutes later we heard the thud of its bombs dropped on Newhaven. The ricks burnt themselves out, the fire-engine returned to Lewes, and the village of Rodmell retired to bed.

A few nights after this I was patrolling the village about three o'clock in the morning and the siren sounded an air-raid warning. Some time later a German bomber flew over the village, and when it was almost immediately overhead there was a tremendous crash about a hundred yards ahead of me. I rushed to the spot, which was a large building at the top of the village. This strange building had been built by a speculative builder as a warehouse in the middle of the nineteenth century when it was thought that the railway from Lewes to Newhaven, shortly to be constructed, would run through the village. The speculation missed fire when the railway was built on the other side of the river. The warehouse remained empty and derelict for years, but after the 1914 war was converted into flats. I expected to find the building half-demolished, but I could find no sign of anything wrong with it. As I walked round it a head appeared out of a window and a female voice said: "It's all right, sir. The noise was caused by our cat; he jumped into an empty tin bath which was outside at the top of the steps and he upset it so as it went clattering down with him into the yard."

One more curious Rodmell incident. I was playing a game of bowls with Virginia, her niece, and a young man one summer afternoon. I was just about to play a shot when a plane came overhead and the young man said to me: "That's an odd plane; what is it!" I looked up and there very low down flying slowly was an old-fashioned-looking plane. I was more interested in my shot, and said casually: "O, that must be a Lysander". I played my shot, and as I did so there was the noise of splattering on the leaves of the great chestnut tree near by and I realized at once that it was the noise of bullets. I looked up again at the plane, which was circling slowly over the garden, and saw the swastika on it. The bullets did not come from the plane, but from a searchlight unit on the down above the village. We lay on the grass while more bullets spattered overhead. The plane circled round and then very very slowly flew off across the water-meadows. As we watched it, it came down in the brooks about two miles away. We heard later that the German pilot was a mere boy; he was taking mail in an antediluvian plane from France to the Germans in the Channel Islands. Losing his way, he had mistaken Sussex for Jersey. Some of the bullets hit his plane, but he himself was unwounded.

When the Battle of Britain and the bombing of London in earnest began, one watched daily in Rodmell the sinister preliminaries to destruction. First the wail of the sirens; then the drone of the German planes flying in from the sea, usually to the east of Rodmell and Lewes. On a clear fine day one could see the Germans high up in the sky and sometimes the British planes going up to meet them north of Lewes. There was very little fighting in the air immediately over the Ouse valley for the Germans flew regularly in a corridor more to the east. I saw only two incidents. A British Hurricane was shot down in a field near the river and I saw a German plane shot down into the water-meadows near Lewes. In the latter incident I felt strongly the spectacular or visual unreality—and also it must be admitted a strange beauty—in modern warfare. It was again on a lovely, windless, sunny autumn afternoon and there was considerable air activity all round us during a raid. I walked out on to the lawn overlooking the water-meadows to watch what was happening. Suddenly a German plane came flying low over the downs towards Lewes pursued by a British fighter. The fighter gained on the German and there was the sound of firing. Suddenly the German plane went straight up into the air, turned over upside down in a great slow loop, and, as the British plane shot by underneath it, fell into the meadows. A column of smoke rose absolutely straight into the windless sky.

When the night bombings of London began we were still living during the week in Mecklenburgh Square; but when the house was made uninhabitable by bombs, which fell in the square and in the street behind it, we had to live in Rodmell. We were in Mecklenburgh Square in one of the earliest night raids. We did not go into the shelter which had been built in the centre of the square, though we had a look at it. We thought it better to die, if that were to be our fate, in our beds. There was a good deal of noise, but nothing that night fell very near to us. I wanted to see Kingsley Martin about something and at ten o'clock next morning Virginia and I walked round to the *New Statesman* office in Great Turnstile just south of Holborn. There was no sign of the blitz until we turned out of Gray's Inn into Holborn. It was the first sight we had of the aftermath desolation of an air raid. One got accustomed, or almost accustomed, to seeing it, as the war went on; but seen for the first time, it had a strange impact on one. The first thing one noticed was the litter of glass from the windows all the way up Holborn. Then the silence—no traffic and very few people about. The holes in the façade of the street where a building had been destroyed, like a tooth missing in a mouth. A smell of burning. We went down the narrow alley and into the *New Statesman* building. It seemed to be entirely empty. My recollection is that it was intact but there was a good deal of water about. I went along to Kingsley's room and there standing by himself with his hat in his hand in the empty room was a man who said to me in a plaintive voice: "I have come to see Mr Kingsley Martin". His look and tone of voice seemed to imply that I had caused the desolation and the absence of Mr Kingsley Martin.

"So have I," I said, "but he does not appear to be here." He gave me a nasty look and left the room.

In the latter part of 1940, when we were living permanently in Rodmell, it was during the mass raids on London a sinister thing to listen every night to the drone of the German planes flying inland. In the hour or more of silence which followed, it was horrible to know that London was being bombed and burnt. Then the silence was broken by the drone of the returning planes. There was one strange incident connected with these night raids. Very often after the raiders had flown in to London, it seemed as if one plane had been left behind to fly round and round the Ouse valley. It did this until just before the other German bombers returned, when it dropped bombs, as it seemed, erratically anywhere in the water-meadows or on the downs. The curious manœuvres of this lone wolf went on for some time and then suddenly stopped. I knew a stonemason in Lewes who, during the war, was one of those monitors of the skies who kept a check on and reported all planes entering the section of heaven for which he was responsible. When I met him one day, some time after the German raids had stopped, I asked him whether they had noticed this lone wolf of a bomber and whether he had any explanation of its curious behaviour. He said that the spotters were convinced that the lone wolf was a coward who habitually dropped out of his squadron in order to avoid the terrors which awaited him over London. They believed that his sudden disappearance was due to his having been "executed" by his fellow-airmen; they probably shot him down into the Channel on their way back one night.

I must now leave these tales of the war and return to our daily life in 1939 and 1940. All through the last months of 1939 and the first of 1940 we divided our time between Mecklenburgh Square and Monks House. Virginia was working very hard—too hard in fact. She had begun both *Roger Fry* and *Between the Acts* in the first half of 1938 and was still writing them all through 1939. She enjoyed writing *Between the Acts*, but the life of Roger became a burden to her. It was a book which I do not think she ought to have undertaken, but she had been overpersuaded by Margery Fry to do it. The orderly presentation of reality, which remorselessly imposes its iron pattern upon the writer who rashly tries to discern and describe it in the infinite kaleidoscope of facts, was not natural to Virginia's mind or method. Four times in her life she forced herself to write a book against her artistic and psychological grain; four times the result was bad for the book and twice it was bad for herself. She said herself that she wrote *Night and Day* as a kind of exercise; she had a shadowy *Jacob's Room* in her mind, a novel which would break the traditional mould of the English novel, which would have a new form and method, because she felt that the traditional form and method did not allow her to say what she wanted to say. Before she broke the mould, she thought she ought to prove that she could write a novel classically in the traditional mould.

The second time was *The Years*. In 1932 after *Orlando* and *Flush* she decided to write a "family novel", a form of fiction popular at the time. When she began to write it in 1932 with the title *The Pargiters* (which marked it as a family novel) she wrote in her diary:

It's to be an Essay-Novel, called *The Pargiters*—and it's to take in everything, sex, education, life etc.

and she added:

> What has happened of course is that after abstaining
> from the novel of fact all these years—since 1919—and
> *N. & D.* is dead—I find myself infinitely delighting in
> facts for a change, and in possession of quantities be-
> yond counting: though I feel now and then the tug to
> vision, but resist it. This is the true line, I am sure, after
> *The Waves—The Pargiters*—this is what leads naturally
> on to the next stages—the essay novel.

There was another thing which perhaps to some extent
influenced her in this determination to write "a novel of
fact". She was oversensitive to any criticism, and one of
the things most often said against her as novelist was that
she could not create real characters or the reality of everyday life. I think that in 1932, beginning *The Years*, at the
back of her mind was the desire or determination to prove
these critics wrong.

The third time was the biography of Roger Fry, which
she began to write six years after she had begun *The Years*.
Being a biography, it was far more concerned with the
facts and determined by facts than a novel—it was fact,
not fiction. When I first read it I thought there was a flaw
in it, and as Virginia recorded in her diary on March 20,
1940,* walking over the water-meadows I tried—no
doubt too emphatically—to explain to her what I felt
about it. Like everything of hers, it had things in it which
could only have been hers and very good they were, but
the two parts of the book did not artistically fit together
and she allowed the facts to control her too compulsively

* *A Writer's Diary*, p. 316.

so that the book was slightly broken-backed and never came alive as a whole. Roger's sister Margery, Vanessa, and his friends and relations generally disagreed with me, but I still think I was right. There is something a little dead about *Roger Fry: A Biography*. Virginia herself, in the passage from her diary quoted above, in 1932 said that *Night and Day* was dead. In the third factual book, *The Years*, there is again something radically wrong. The "factual" novel, the facts, got out of control, the book became inordinately long and loose. I have described in *Downhill All the Way** the terrible time we had when *The Years* was finished and Virginia was faced with the proofs. She tried to regain control of the facts and the novel, to eliminate the length and the looseness, by drastic cutting, but the operation was not really successful, the Bed of Procrustes was unable to turn the book into a masterpiece; it is a psychological and familial chronicle rather than a history, and, like *Night and Day* and *Roger Fry*, was slightly dead even at the moment of birth.

Virginia was an intellectual in every sense of the word; she had the strong, logical, down to earth brain which was characteristic of so many of her male Stephen relations: her grandfather, James Stephen of the Colonial Office; her father, the author of *History of English Thought in the* 18*th Century* and *Hours in a Library;* her uncle, James Fitzjames, the High Court judge; and her brother Thoby. It was not that she could not handle facts or had the weakness in induction or deduction which men so often, without much evidence, consider peculiarly feminine. Her reviews, and indeed these factual books themselves, show

* Pages 153-6.

this. But she could deal with facts and arguments on the scale of a full-length book only by writing against the grain, by continually repressing something which was natural and necessary to her peculiar genius—she shows this in the passage quoted above when she writes: "I feel now and then the tug of vision, but resist it". The result is a certain laboriousness and deadness so different from the quicksilver intensity of the novels in which the tug of vision was not resisted. It is even easier to see the difference if one compares *A Room of One's Own* with *Three Guineas*, the fourth of her factual books. In *A Room of One's Own* there is no shirking of facts and arguments, but they are subjected to and illuminated by vision, and the book tingles with life; *Three Guineas*, in comparison, is oppressed by the weight of its facts and arguments.

In all these cases this forcing herself to write against the grain, to resist her own genius, added to the mental and physical strain of writing a book, the exhaustion and depression which nearly always overwhelmed her when the umbilical cord was severed and the MS sent to the printer. The most desperate example of this was what happened between the writing and the printing of *The Years*.* There is no doubt that the biography of Roger exhausted her in much the same way. She finally finished the book (there was nearly always a final finish with her books) on April 9 and it was at last off her hands with the corrected proofs returned to the printers on May 13. But it had been a terrible grind in many ways. She said that she had "written every page—certainly the last—ten or fifteen times over". At first, with the relief of getting the book off her hands and being able to go back to fiction

* See *Downhill All the Way*, pp. 153-7.

and vision in *Between the Acts*, she seemed to be less worried and more optimistic than she usually was when her book was on the threshold of publication.

The umbilical cord which had bound *Roger Fry: A Biography* to Virginia's brain for two years was, as I said, finally cut when she returned the proofs to the printers on May 13, 1940; 319 days later on March 28, 1941, she committed suicide by drowning herself in the River Ouse. Those 319 days of headlong and yet slow-moving catastrophe were the most terrible and agonizing days of my life. The world of my private life and of English history and of the bricks and mortar of London disintegrated. To drag the memory of them out of one's memory, as I must do now if I am to continue publicly to remember, is difficult and painful. The reluctant recollection of protracted pain is peculiarly painful. The excitement in the moment of catastrophe, the day-to-day, hour-to-hour, minute-to-minute stimulus of having to act, produce an infallible anodyne for misery. I am always astonished to find that one instantly becomes oblivious of the most acute pain if one has to concentrate on anything else, even on a triviality. While one concentrates on crossing a crowded street, in London, the consciousness of the torture of toothache or of being crossed in love is completely obliterated. But there are no distractions or alleviation in one's recollection of misery.

Virginia's loss of control over her mind, the depression and despair which ended in her death, began only a month or two before her suicide. Though the strains and stresses of life in London and Sussex in the eight months between April 1940 and January 1941 were for her, as for everyone living in that tormented area, terrific, she

Monks House, Rodmell: entrance

Monks House: sitting room

Gisèle Freund

was happier for the most part and her mind more tranquil than usual. The entry in her diary for May 13, already published in *A Writer's Diary*,* gives the atmosphere of those violent days and the ambivalence of her mood and mind so vividly that I must quote it here.

"I admit to some content, some closing of a chapter and peace that comes with it, from posting my proofs today. I admit—because we're in the third day of 'the greatest battle in history'. It began (here) with the 8 o'clock wireless announcing as I lay half asleep the invasion of Holland and Belgium. The third day of the Battle of Waterloo. Apple blossom snowing the garden. A bowl lost in the pond. Churchill exhorting all men to stand together. 'I have nothing to offer but blood and tears and sweat.' These vast formless shapes further circulate. They aren't substances: but they make everything else minute. Duncan saw an air battle over Charleston—a silver pencil and a puff of smoke. Percy has seen the wounded arriving in their boots. So my little moment of peace comes in a yawning hollow. But though L. says he has petrol in the garage for suicide should Hitler win, we go on. It's the vastness, and the smallness, that makes this possible. So intense are my feelings (about *Roger*); yet the circumference (the war) seems to make a hoop round them. No, I can't get the odd incongruity of feeling intensely and at the same time knowing that there's no importance in that feeling. Or is there, as I sometimes think, more importance than ever? I made buns for tea today—a sign my thraldom to proofs (galleys) is over."

*Pages 319-20.

A few days later when there was an appeal to join the Home Guard "against parachutists" I said that I would join, and there was rather "an acid conversation" because Virginia, though she saw that I was "evidently relieved by the chance of doing something", was against it, feeling "gun and uniform to me slightly ridiculous". She admitted that her nerves were harassed under the strain of looming uncertainty. But we discussed again calmly what we should do if Hitler landed. The least that I could look forward to as a Jew, we knew, would be to be "beaten up". We agreed that, if the time came, there would be no point in waiting; we would shut the garage door and commit suicide. "No", Virginia wrote, "I don't want the garage to see the end of me. I've a wish for ten years more, and to write my book, which as usual darts into my brain. . . . Why am I optimistic? Or rather not either way? Because it's all bombast, the war. One old lady pinning on her cap has more reality. So if one dies, it'll be a common sense, dull end—not comparable to a day's walk, and then an evening reading by the fire. . . . Anyhow, it can't last— this intensity, so we think—more than ten days. A fateful book this. Still some blank pages—and what shall I write on the next ten?"

What she wrote was much the same kind of thing which anyone who kept a diary would have to write in the ten days of a great catastrophe of history, the death and destruction of a civilization. In the ten days when heaven and civilization, the country where one was born and bred and lived, one's private life and banking account, are falling about one's head, when one is contemplating suicide by asphyxiation in a damp and dirty garage after breakfast, one continues to cook and eat one's eggs and bacon for breakfast. Greeks were doing that, I feel sure, in Thebes and Athens 2,304 years ago when Alexander the Great and his armies were destroying the civilization of Homer, Pindar, Sophocles, and Plato and every house except one in the city of Thebes—it was characteristic of one of the great conquerors, the great destroyers of civilization, the inhuman pests of human history, that Alexander

> "The great Emathian conqueror bid spare
> The house of Pindarus, when temple and tower
> Went to the ground",

and he sold all the inhabitants as slaves. It must have been the same when, 1,558 years ago in Rome the Romans were eating the equivalent of our eggs and bacon and Alaric, the Goth, was on the point of capturing and sacking their city. It must have been the same with the millions of victims of Jenghis Khan, and of the Turks when Constantinople fell in 1453, and of the drums and tramplings of all conquests from the Egyptian and Sumerian to that of Hitler.

Moore and Desmond MacCarthy came and stayed with us for the weekend of May 18. It was the last time that I was to see Moore. Desmond and Moore together, the one talking, talking, the other silent in the armchair, were inextricably a part of my youth, of the entrancing excitement of feeling life open out in one and before one. I could shut my eyes and *feel* myself back in 1903, in Moore's room in the Cloisters of Trinity or the reading parties at the Lizard or Hunters Inn. With the eyes open we were older. I myself either have never grown up or was born old, for I have always had the greatest difficulty in

47

feeling older. My "I", that particle indestructible except by death, which answered to that particle in the drowning puppy, seems to me exactly the same in the child in Lexham Gardens, the undergraduate in Moore's room in Trinity, the middle-aged man of 1940, even the old man writing his memoirs. It is easy to feel the stiffness in one's knees, and it is of course only delusion that one does not feel it in one's heart and brain.

Moore was older. The extraordinary purity and beauty of character and of mind were still there, the strange mixture of innocence and wisdom. The purity, moral and mental, was the most remarkable of Moore's qualities; I have never known anything like it in any other human being. It was as though Socrates, Aristotle, and the Pure Fool—the Reine Tor is it?—had grown inextricably entangled in the same mind and body. Bertrand Russell in the second volume of his Autobiography writes of that other Cambridge philosopher Wittgenstein: "He was perhaps the most perfect example of genius as traditionally conceived, passionate, profound, intense and dominating. He had a kind of purity which I have never known equalled except by G. E. Moore." There was a streak of aggressive cruelty in Wittgenstein—I once saw him so brutally rude to Lydia Keynes at luncheon, when he was staying with Maynard, that she burst into tears. Moore had the passion, the profundity, the intensity, and the purity, but he seemed to be completely without cruelty and aggression. Age had blunted the passion and softened the intensity. It had chiselled away some of the beauty of his face. He told me that a little time before his visit to us he had a strange and alarming black-out which, he felt, had left its mark on him. But the luminous purity of mind and spirit was in him at the age of sixty-seven in Rodmell just as it was forty years before when I first saw him in Trinity.

Age had left its mark too upon Desmond; both his face and his mind were age-beaten. Asthma and the erosion by the incessant little worries of life had to some extent dimmed and depressed him, but every now and again memory, which could make an artist of Desmond, and his amused devotion and affection for Moore, inspired him so that the years fell away and one again felt to the full the charm of his character and his conversation.

There we sat in May 1940, Moore, Desmond, Virginia, and I in the house and under a hot sun and brilliant sky in the garden, in a cocoon of friendship and nostalgic memories. At the same time the whole weekend was dominated by a consciousness that our little private world was menaced by destruction, by oecumenical catastrophe now beginning across the Channel in France. It was, of course, a week before the capitulation of the Belgians, but the German offensive had been in operation for ten days, the tension was unrelieved and one's memories of unremitting defeat in the terrible first years of the 1914 war left one inevitably with premonition of disaster, so that even hope was a kind of self-indulgence and self-deception. The restless foreboding of those days was broken for me by a strange, grim incident which I always in memory connect with them. There lived in Rodmell at that time a working-class man whom I will call Mr X. I knew the whole family very well. The youngest son, now about eight or nine years old, had been injured at birth and was completely "retarded". He could not speak or feed himself and could hardly walk. As is so often the case, his mother adored him and devoted her life to looking after him, which indeed was a full-time job. Some years before the eldest son had come to me and asked me to try to persuade Mrs X to send the boy to a mental home, as he and the rest of the family thought that his mother was ruining her life by immuring herself with the child. They had tried to argue with her, but she would not listen to them; he thought that she might listen to me. She did listen to me, but what I said had no effect upon her.

The eldest son, Percy, was in 1940 called up for service in the army, and one evening he came to see me. He said his regiment was leaving in a day or two for France, and he was very much worried about his mother, who was really destroying her health and all happiness by her devotion to the youngest son. He would embark for France much relieved if I would make another attempt to persuade Mrs X to send the child to a home. This time I was successful and I went to the Medical Officer, who already knew about the case, and asked him to get the boy into a home. He did so, and at first everything went well; but after about two weeks Mrs X came to me and said that the boy was being starved and ill-treated, was getting very ill, and must be given back to them. Then one morning Mr and Mrs X appeared in my garden dressed in their Sunday clothes. They had hired a taxi and asked me to accompany them to the Medical Officer and demand the child.

There followed some painful hours. I agreed to go to the M.O. provided that they left the business to me and did not start abusing him and the Home for starving the boy. They promised, but within five minutes of our being shown into the M.O.'s room Mrs X was making the wildest accusations against him, the Home, and the nurses. The M.O. behaved admirably; he rang up the Home and arranged that if we went there immediately, the boy would be handed over to us. I do not think that I have ever had a more unpleasant pilgrimage in my life than that to the Home and back to Rodmell, sitting in the taxi with the unfortunate parents. The boy was delivered to us wrapped in blankets. He was obviously ill and a week or ten days later he died. There was an inquest, at which Mrs X repeated her accusations against the nurses and everyone connected with the Home, but the verdict was death from natural causes.

This kind of tragedy, essentially terrible, but in detail often grotesque and even ridiculous, is not uncommon in village life. At the time its impact upon me was strong and strange; somehow or other it seemed sardonically to fit into the pattern of a private and public world threatened with destruction. The passionate devotion of mothers to imbecile children, which was the pivot of this distressing incident, always seems to me a strange and even disturbing phenomenon. I can see and sympathize with the appeal of helplessness and vulnerability in a very young living creature—I have felt it myself in the case of an infant puppy, kitten, leopard, and even the much less attractive and more savage human baby. In all these cases, apart from the appeal of helplessness, there is the appeal of physical beauty; I always remember the extraordinary beauty of the little leopard cub which I had in Ceylon, so young that his legs wobbled a little under him as he began jerkily to gambol down the verandah and yet showing already under his lovely, shining coat the potential rippling strength of his muscles. But there is something

51

horrible and repulsive in the slobbering imbecility of a human being. Is the exaggerated devotion of the mother to this child, which nearly always seems to be far greater than her devotion to her normal, attractive children, partly determined by an unconscious sense of guilt and desire to vindicate herself and her child? It is rather strange that I have twice been asked to interfere in a case of this sort. I once had a secretary who came to me and asked me to speak to her sister just as Percy had asked me to speak to his mother. The sister had a very highly paid post as buyer of dresses for one of the large Oxford Street stores. She had an imbecile son and, like Mrs X, devoted the whole of her domestic life to him. He was fifteen, physically strong, and occasionally violent. She refused to send him to a Home and I was asked to try to persuade her to do so. I failed, but in this case not long afterwards the boy's behaviour became so alarming that it was no longer possible for his mother to keep him at home.

To return to our daily life in May and June 1940, before the bombing of London began, it slipped into a regular routine. Virginia, having got rid of *Roger Fry*, settled down to writing *Between the Acts* or *Pointz Hall* as she still called it. It went well and on the whole she was quite happy about it. By May 31 she was within sight of the end, for she had already written the passage about "scraps, orts and fragments", which in the printed book is only thirty-five pages from the end. In order to give her uninterrupted quiet for writing the novel, we divided our time between Rodmell and London. Every two weeks we went up to Mecklenburgh Square and stayed there for four days. That ensured Virginia ten days out of every fourteen in which she could write uninterruptedly in Rodmell.

The four days in London were always pretty hectic. My main business was The Hogarth Press. John Lehmann had come into the Press as a partner in 1938 and he was, to all intents and purposes, a managing director. It was a difficult time already for the Press, and it became much worse when the blitz started. In order to relieve the strain on the staff, we had them one by one to spend a weekend with us at Rodmell. In my bi-weekly four days in Mecklenburgh Square, in addition to the hours which I spent with John over The Hogarth Press business, I used to go to the House of Commons for the Labour Party Advisory Committees, of which I was still secretary, and to the Fabian Society where I was still a member of the Executive Committee and chairman of the International Bureau. We also crammed as much social life as possible into our four days, having many of our friends to dinner. For instance, in the two London visits of May 21-24 and June 4-7 we saw T. S. Eliot, Koteliansky, William Plomer, Sybil Colefax, Morgan Forster, Raymond Mortimer, Stephen Spender, Kingsley Martin, Rose Macaulay, and Willie Robson.

There was in those days an ominous and threatening unreality, a feeling that one was living in a bad dream and that one was on the point of waking up from this horrible unreality into a still more horrible reality. During the months of the phoney war, when everything seemed to be for the moment inexplicably suspended, there was this incessant feeling of unreality and impending disaster, but it became even stronger during the five weeks between Hitler's invasion of Holland and Belgium and the collapse of France. There was a curious atmosphere of quiet fatalism, of waiting for the inevitable and the aura of it still lingers in the account of our days in London which Virginia gives in her diary. For instance, in the first week in June, with the great battle raging in France, Virginia and I with Rose Macaulay and Kingsley Martin sat talking after dinner until 2.30 in the morning. Kingsley, "diffusing his soft charcoal gloom", prophesied the defeat of the French and the invasion of Britain within five weeks. A Fifth Column would get to work; the Government would move off to Canada leaving us to a German Pro-Consul, a concentration camp, or suicide. We discussed suicide while the electric light gradually faded and finally left us sitting in complete darkness.

Then quickly came the collapse of the French and the retreat of the British to Dunkirk. In Rodmell Dunkirk was a harrowing business. There was not merely the public catastrophe, the terrible suspense with Britain on the razor's edge of complete disaster; in the village we were domestically on the beaches. For Percy, who had come to me about his mother only a few days ago, and Jim and Dick and Chris, whom I had known as small boys in the village school and watched grow up into farm workers and tractor drivers, were now, one knew, retreating like the two grenadiers of the Napoleonic wars, driven back to the Dunkirk beaches. There presumably they were waiting and we in Rodmell waited. On June 17 Percy suddenly appeared in the village; his story was the soldier's story which, with an infinite series of variations, has been told again and again ever since, and before, Othello "spake . . . of hairbreadth 'scapes i' the imminent deadly breach". It was not the less moving—partly because Percy was still badly shaken by his experiences; everything was still shockingly vivid to him, and it eased his mind somewhat

to describe it vividly to someone else. He described the
retreat through Belgium, the Belgian woman who gave
him bread for nothing—"no Frenchwoman would do
that—a nasty gabbling, panicky lot the French"—an
abandoned jeweller's shop which they looted—he took
some rings and two watches—the rings fell out of his
pocket in the sea at Dunkirk, but he pinned the watches
to the inside of his coat and got them safely back to
England—they were strafed incessantly by German planes
on the beaches and "the bullets were like moth-holes in
my coat"; there were no British planes; when they had
to pass a pill-box, the officer made the men take off their
boots and crawl past it and he then went himself and
flung a grenade into it—he saw his cousin lying dead on
the beach and also another Rodmell chap who used to
live up the village street—on the beach a chap showed
him a silk handkerchief which he had bought for his "joy
lady" and, as he did so, a bomb fell and killed the man.
Percy still has the silk handkerchief—he swam out to sea
to a small boat, *The Linnet*; when he got near it, they
shouted to him: "Say, chum, can you row?" and when
he said that he could, they pulled him into the boat—they
rowed for five hours before they saw the English coast,
and they were so exhausted that, when they at last landed,
they did not know or ask where they were and they did not
know whether it was night or day. Actually he landed at
Ramsgate, got lifts to somewhere on the Eastbourne road,
and walked through the night to his sister's cottage in
Rodmell. She, in that early morning, looking out of her
kitchen window, saw a soldier whom she did not recognize
at first, hatless, his tunic bloody and full of holes, his boots
in rags, lying exhausted outside the front door. His

morale was pretty low; he thought we were beaten—we had no arms or planes; yet he soon recovered, returned to his regiment, and fought until the end of the war, ending up in the army of occupation in Germany and marrying a very nice German girl from Hamburg.

Friday, June 14, the Germans took Paris, and we spent the day incongruously—or from another point of view appropriately—with history being made so catastrophically across the Channel—for we took a journey from the present into the past. Virginia had never been to Penshurst, the magnificent Elizabethan mansion in Kent which has remained in the Sidney family ever since the time of the romantic Sir Philip Sidney, who wrote *Arcadia* and was killed at the battle of Zutphen in Holland nearly four hundred years ago. We had more than once been on the point of making an expedition with Vita Sackville-West to see the house on a day when it was open to the public. And this is what we did on Friday, June 14. Penshurst is an epitome of English history with its vast banqueting hall and its vast clutter of pictures, furniture, utensils—some of them beautiful, but many appallingly ugly—which the great aristocratic families accumulate in their castles and mansions through the centuries. It was amusing to visit Penshurst with Vita who had the blood of all these owners of Elizabethan castles—including probably the Sidneys'—in her veins and seemed always to have a castle or two of her own hanging about her. We went through the rooms and the bric-à-brac, all of which, no doubt, is well worth seeing, but I left with that rather oppressive feeling of passing through centuries of history mummified or more modernly preserved in deep freeze. When we came out from the immense banqueting hall into the open air, Vita said that she must go round and see the owner of Penshurst, Lord de L'Isle and Dudley, for he would never forgive her if he found out that she had been to Penshurst without seeing him. She rang the bell and was admitted through the front door of what seemed to be a small villa-ish annexe to the great house. Some minutes later she came out and said that Lord de l'Isle insisted that we too must come in. I have rarely felt anything to be stranger and more incongruous than the spectacle of the heir of all the Sidneys sitting in his ancestral mansion. He was an elderly gentleman, obviously in poor health, sitting in a small ugly room. Whether my scale of values had been distorted by looking at all the "treasures" accumulated over four centuries in the great house, I do not know, but Lord de L'Isle and Dudley seemed to me to be having his tea in a room furnished by Woolworth, the only luxury being a small bookcase full of Penguins. He was a nice, but melancholy man, complaining to Vita that he sat in his room most of the time and saw very few people, his only distraction being an occasional visit to Tunbridge Wells for a rubber of bridge.

There was something historically absurd and touching, ironically incongruous and yet, in that particular moment of history, appropriate in the spectacle of Vita and Lord de L'Isle and Virginia and me sitting together in that ugly little room. Vita was in many ways an extremely unassuming and modest person, but below the surface, and not so very much below, she had the instinctive arrogance of the aristocrat of the ancient regime, and above the surface she was keenly conscious of the long line of her ancestors, the Sackvilles, Buckhursts, and Dorsets, and of the great house of Knole a few miles away in Kent. Her

ancestor, Thomas Sackville, Lord Buckhurst and Earl of
Dorset, Lord High Treasurer of England, might well
have driven over from Knole to visit Lord de L'Isle's
ancestor, Sir Philip Sidney, at Penshurst four hundred
years ago. He would not have taken either Virginia's or
my ancestors with him, for Virginia's ancestors were
labouring as little better than serfs in Aberdeenshire and
mine were living "despised and rejected" in some conti-
nental ghetto. Thomas Sackville would have been received
in state in the enormous banqueting hall, with the great
fire burning in the centre of the hall, and he would have
been given a mighty feast on plates of gold or silver, and
wine or mead in golden cups from silver flagons. In 1940
the descendants of the Scottish serf and the ghetto Jew,
on payment of 2s. 6d. each, visited the banqueting hall
and the sitting-rooms and bedrooms, with their accumu-
lation of ancient furniture, pictures, silver, and china
(which must now be worth hundreds of thousands of
pounds), while Lord de L'Isle, the owner, and the descen-
dant of Thomas Sackville, sat in a poky little room drinking tea from rather dreary china. I felt that in that room
history had fallen about the ears of the Sidneys and the
Leicesters, the Sackvilles and the Dorsets, while outside,
across the Channel, in France, history was falling about
the ears of us all.

The collapse of our world continued three days later
when the Pétain Government in France asked for an
armistice with the Germans and on July 4 came the harrowing incident of our attack upon the French fleet. It was
then, however, that we, like so many other people, had
that strange sense of relief—almost of exhilaration—at
being alone, "shut of" all encumbrances, including our

allies—"now we can go it alone", in our muddled, make-shift, empirical English way. Then the bombing began. My first experience of public behaviour during an air raid was in the House of Commons. I was in a Committee Room attending a meeting of the Labour Party Advisory Committee when the sirens started. We all had to go down into the basement, ministers, M.P.s, officials, cleaners, and general public. For a quarter of an hour we stood or sat about rather solemn and self-conscious. One soon learned not to be solemn and self-conscious in the blitz. I hated air-raid shelters and only once went into one at night, the shelter constructed in Mecklenburgh Square, during a particularly heavy bombing. I hated the stuffi-ness and smell of human beings, and, if a bomb was going to get me, I preferred to die a solitary death above ground and in the open air. Like so many convinced and fervent democrats, in practice I have never found human beings physically in the mass at all attractive—there is a good deal to be said for solitude whether in life or in death. When death comes, I should choose, as some wild animals do, to go off and meet it alone—but not in an air-raid shelter or underground. I think that I felt the physical oppressive-ness of human beings in the mass most heavily when, in the worst days of the blitz, I passed through Russell Square Underground Station at night on my way to or from my house in Mecklenburgh Square. It was used, like the other Underground Stations, as a dormitory and air-raid shelter by dozens of men, women, and children, on mattresses wrapped in sheets and blankets and lying side by side all the way down the platform as if they were sardines in a gigantic tin.

One missed something, of course, by not congregating

with one's fellows when the bombs fell. Everyone felt the extraordinary blossoming of the sense of comradeship and good-will which settled upon us in London during the blitz, and the falling of bombs loosened our tongues. Queer little scenes and conversations, under those conditions, engraved themselves upon one's memory. For instance, one day, driving up from Rodmell to Mecklenburgh Square, we ran into a nasty air raid in Wimbledon. When we reached the Common, the bombs began to fall unpleasantly near us, and when I saw a "pillbox" not far from the road, we left the car and took shelter in it. It was the usual large, square room with a cement floor and slits for guns in the thick walls. It was already inhabited. In one corner was a young woman typist who had been bombed out of her lodging and was now "temporarily" living in the pillbox with a small suitcase containing all her possessions. In the other corner was a family consisting of husband, wife, and child. They had a camp bed, two chairs, several boxes, pots and pans, china and cutlery, and a Primus stove. They were having a cup of tea and we and the young woman were invited to join them. In two minutes we were all chatting happily like old friends. The man was a printer, originally from the North of England. Three weeks before he had taken a job with a printing firm in Wimbledon and with great difficulty had managed to find himself a small house. The first night he moved in a bomb fell in front of the house, blew out all the windows and blew off half the roof. The family was intact, but he could not find a lodging, so he moved into the pillbox. He was a typical printer, and printers, I had learned by experience, are or were typical of the working-class élite, the trade unionist, skilled worker. A year before

he had been living a stuffy, petit bourgeois* life behind
lace curtains in some dreary, respectable back street. The
war and the bombs seemed completely to have changed
his outlook on life. He had not quite reached the stage of
wisdom in the Old Testament's "Let us crown ourselves
with rosebuds, before they be withered" and "Let us eat
and drink; for tomorrow we shall die", but he seemed to
have accepted Christ's recommendation in the New
Testament: "Take therefore no thought for the morrow:
for the morrow shall take thought for the things of itself.
Sufficient unto the day is the evil thereof." And so
deserting the best parlour and the lace curtains, he was
living, more or less contentedly, a kind of nomad existence
in a machine-gun emplacement on Wimbledon Common.

Early in August the German mass raids began on
London, and almost every night their bombers roared over
our heads at Rodmell on the way to London. I still had to
pay the rent of 52 Tavistock Square; there was now no

* Karl Marx was hopelessly wrong in believing that the bourgeoisie would be destroyed by the proletariat. On the contrary, as in so
many other fields, in the class struggle and the world of economics
the victors are absorbed and swallowed up by the vanquished as if
they had fallen into a social jelly or quicksand. As soon as the
workers lose their chains they adopt the food, clothes, habits,
mentality, ambitions, and ideals of the middle classes. From Russia
downwards there is no dictatorship of the proletariat, but always
merely a dictatorship of the bourgeoisie under another name. Engels
was the only leading Marxist who saw this and had the honesty to
admit it (up to a point) when he wrote: "The English proletariat is
becoming more and more bourgeois, so that the most bourgeois of all
nations is apparently aiming ultimately at the possession of a
bourgeois aristocracy and a bourgeois proletariat *as well as* a
bourgeoisie. For a nation which exploits the whole world this is, of
course, to a certain extent justifiable."

chance of anyone taking over the last year of my lease, for many people were leaving London in order to escape from the blitz. I was in correspondence with the Bedford Estate: I had asked them whether they would remit or reduce my rent as the house was empty and I had moved to Mecklenburgh Square. Then one day I received a letter from the Estate saying that the matter had now been settled, for the house had been completely destroyed the previous night. The next time I was in London, I went round to see the ruins of the house which we had lived in for fifteen years. It was a curious and ironic sight, for on the vast conical heap of dust and bricks precisely and meticulously perched upright upon the summit was a wicker chair which had been forgotten in one of the upper rooms. Nothing beside remained except a broken mantel-piece against the bare wall of the next-door house and above it intact one of Duncan Grant's decorations.

On September 10 I drove up to London, but found that it was impossible to get into the house in Mecklenburgh Square. The police had cordoned off the Square after evacuating the inhabitants. The neighbourhood had been badly bombed the night before and there was an un-exploded bomb in the ground in front of our house. The Hogarth Press had come to a standstill. There was nothing to do but drive back to Rodmell and wait until they exploded the bomb. But three days later I drove up for the day again in order to meet John Lehmann and discuss what we should do about the Press. The bomb had been exploded and our house was in a dreadful state, all the windows blown out, doors hanging on one hinge, and the roof damaged. It was soon still further wrecked by the terrible havoc caused by a land mine which fell at the back, killing several families and blasting all our rooms in reverse direction from the previous bombing.

The Hogarth Press premises in the basement and our flat on the third and fourth floors were uninhabitable. All the windows had been blown out; most of the ceilings had been blown down, so that, in most places, you could stand on the ground floor and look up with uninterrupted view to the roof while sparrows scrabbled about on the joists of what had been a ceiling; bookcases had been blown off the walls and the books lay in enormous mounds on the floors covered with rubble and plaster. In the Press books, files, paper, the printing-machine and the type were in a horrible grimy mess. The roof had been so badly damaged that in several places it let the rain in and the water-pipes in the house had been so shaken by the blast that occasionally one burst without warning and sent a waterfall down the stairs from the third floor to the ground floor.

In those days the Press printed many of its books with the Garden City Press, Letchworth, Herts., and they nobly came to our rescue. They offered to give us office accommodation within their printing works for our staff if we evacuated them to Letchworth. We accepted gratefully and for the remainder of the war the entire business of The Hogarth Press was carried on from Letchworth; all our staff agreed to stay with us and migrate to Hertfordshire. The next five years they spent uncomplainingly in lodgings in a strange town away from their homes and friends. John Lehmann continued to live in London and his mother's house on the Thames, continually travelling to Letchworth to supervise and manage the publishing business. I only occasionally took the long, tedious (in wartime) journey from Lewes to Letchworth and back in the day. On one of these journeys I saw the grimmest London devastation of the blitz. I caught an early train from Lewes to London Bridge. When I walked out of the station I found that half the city had been destroyed during the night. There was no traffic, no buses or taxis. I started to walk to King's Cross and got to Cannon Street all right, but as soon as I started to walk north from Cannon Street, though I knew every street there, I completely lost my way. Half the streets had disappeared into smouldering heaps of rubble and were unidentifiable. What was most extraordinary and sinister was the silence. There was no traffic, since most of the streets were blocked with the debris of buildings; there were hardly any pedestrians. There were many fire-engines and firemen still playing their hoses on burning ruins and every now and again I met a policeman. A pall of smoke hung just above one's head and everywhere there was an acrid smell of burning. Every now and again, too, through the smoke and above the ruins I caught a glimpse of St Paul's, and, though half the time I did not know exactly in what street I was, I steered a roughly north-western course by the Cathedral and so eventually found myself in the Farringdon Road and so by comparatively unblitzed streets to King's Cross.

Having lived through the two world wars of 1914 and 1939, I can say that my chief recollection of war is its intolerable boredom. In this respect the first was, I think, worse than the second, but the second was bad enough. It took me about seventeen hours to get from Rodmell to Letchworth and back, and owing to the bombing one was always waiting hours for trains which never came or in trains which could not move because a bomb had fallen on the line ahead of one. More than once I have sat in the train for four hours or more on the journey from Victoria to Lewes which normally took an hour; the main line had been bombed and we had to trickle half-way round Surrey and Sussex in order eventually to reach Lewes via Horsham and Brighton. And having eventually reached Lewes bored and hungry, one found the town in pitch darkness, the last bus gone, no taxis, and nothing for it but to face a four-mile walk in the rain. If I ever prayed, I would pray to be delivered, not so much from battle, murder, and sudden death, but rather from the boredom of war.

Having settled The Hogarth Press in the Letchworth Garden City Press, we then had to consider what to do with our private goods and chattels scattered over the floors of the rooms in Mecklenburgh Square. Something had to be done to rescue them from the wind and rain which swept over them through the shattered windows. Eventually I managed to arrange for their removal to Rodmell. But there was not only the furniture which filled six or seven rooms; there were thousands of books, a large printing machine, and a considerable quantity of type and printing equipment. I succeeded in renting two rooms in a Rodmell farmhouse and a large storeroom in another Rodmell house. Into these and into every spare space in my own house we stacked the mountains of books, furniture, and the equipment of the kitchen chaotically mixed up with the equipment of the printing room.

In a large ground-floor sitting-room of Monks House the thousands of books which we had had in London were piled on tables, chairs, and all over the floor. I had always had a passion for buying and accumulating books, and so had Virginia, but she had also inherited her father's

65

library. This was the kind of library which, in the spacious and affluent days of Queen Victoria, a distinguished gentleman, who edited the *Dictionary of National Biography* and was an eminent critic and essayist, was apt to acquire. There upon his shelves in complete editions of ten, twenty, thirty, or forty volumes, often pompously bound in calf, stood the rows of English and French classics. Now they were piled into the sitting-room in grimy, hopelessly jumbled heaps. They led one day to one of those trivial, but unexpectedly pleasant, incidents that occasionally mitigated the menace and monotony of the war. In the years before the invasion of France large numbers of troops were being trained in the South of England and there was an unending succession of regiments on troop marches through Rodmell. They often camped in my field and quite often I used to put up the officers in one of the bedrooms and in a garden room. One day in the late summer there was a Lancashire regiment in my field, but I had seen nothing of the officers. In the late afternoon I was on a ladder gathering figs from a large fig-tree when I heard someone call to me. I looked down and there stood a very swarthy subaltern. He wanted to borrow something off me and I took him into the house. The door of the room in which the books were piled was open, and, when he saw them, he rushed into the room in the greatest excitement, seized on a book, and, though it was the *Novum Organum* of Francis Bacon, began to read it. One has learned to expect the vagaries of human beings to be infinitely unpredictable, but I must say that I was slightly surprised to see an unknown Lieutenant in an English regiment become engrossed in the *Novum Organum*. My visitor, who spent practically the whole of the next two days with me, was a very interesting and amusing man. He was a Pole belonging to the landowning class. Before the war he was in the Diplomatic Service, and, when it broke out, he was in the Polish Embassy in Washington. He wanted to fight against the Germans and managed to reach England. I forget how exactly the wheels of Polish politics revolved around Sikorski and Anders, but I think that my Lieutenant did not like them, and so somehow or other he contrived to obtain a commission in an English regiment. He gave one the impression of being what is called a tough customer, and in a tight corner I should have preferred to have him on my side rather than against me. But he also had a passion for literature, learning, books, and reading. He had been unable to read and had hardly seen a book for months, and the sight of the mountain of my books had the same effect upon him as the sudden sight of a spring of pure cold water would have upon a man dying of thirst. For the next forty-eight hours he sat in my house from early morning until midnight, devouring book after book, forgetting the war and his regiment. We had our meals in the kitchen and sat long over them discussing love and life and death and politics and literature. It was one of those rare, unexpectedly pleasant and exhilarating interludes in the claustrophobia of war, this sudden, fleeting appearance in one's life of an entirely sympathetic stranger. He and I, completely different in birth, nationality, education, and experience, had come together from the ends of the earth for a moment of time—forty-eight hours—but, under the shadow of death and disaster, sitting over our meagre war rations, we talked as if we had known each other for a lifetime, finding that the world and the universe presented the same delightful, horrible, and ridiculous face to both of us.

My Lieutenant and his regiment disappeared into the fog of war at the end of the two days and, though I remembered with pleasure his toughness, eagerness, and intelligence, I never expected to see him again. But a year or two after the end of the war, one summer day I was in the garden and there suddenly into it came my Polish Lieutenant, in civilian clothes and accompanied by a very pretty young woman. He had survived the war, and, remembering me and my books, determined to come and see me again if he possibly could. Once more we talked for an hour or two and then once more he disappeared into Europe. I have never seen him again, and I hope that he has survived the peace in Poland as he survived the war.

One other incident connected with the regiments which camped in my field showed how international some of our regiments became in the war. One day when (I think) a Kent regiment was camped in my field I put three or four officers up in a room which contained a large bookcase full of translations of Virginia's books into almost every European language. In the afternoon I went into the room to see whether the officers had everything they needed. I found only one subaltern sitting in a chair and reading a book. To my astonishment the book was a Czech translation of *Flush*. The Lieutenant was a Czech and he had been astonished and delighted to find in a house in Sussex a book in his native language.

I must return to chronology, to the chronological narrative of this autobiography. The return is to August or September of 1940—and perhaps this is the point at which I might for a moment digress about autobiographical digressions, and make, not an excuse or defence, but an explanation. Some critics of the previous volumes of this autobiography have complained of my digressions, my habit of "not sticking to the point", and one or two have politely suggested that the cause is old age, garrulous senility. "A good old man, sir, he will be talking; as they say, 'When the age is in, the wit is out'." I would not deny the explanation or indictment, but I also digress deliberately. Life is not an orderly progression, self-contained like a musical scale or a quadratic equation. For the autobiographer to force his life and his memories of it into a strictly chronological straight line is to distort its shape and fake and falsify his memories. If one is to try to record one's life truthfully, one must aim at getting into the record of it something of the disorderly discontinuity which makes it so absurd, unpredictable, bearable.

Looking back from Virginia's suicide in March 1941 to the last four months of 1940, I have naturally often asked myself why I had no forebodings of the catastrophe until the beginning of 1941. What was the real state of her mind and her health in the autumn and early winter of 1940? I thought at the time and still think that her mind was calmer and more stable, her spirits happier and more serene, than was usual with her. If one is in the exact centre of a cyclone or tornado, one finds oneself in a deathly calm while all round one is the turmoil of roaring wind and wave. It seemed as if in Rodmell in those last months of 1940 we had suddenly entered into the silent, motionless centre of the hurricane of war. It was a pause, only a pause, as we waited for the next catastrophe; but we waited in complete calm, without tension, with the threat of invasion above our heads and the bombs and bombing all round us. It was partly that we felt physically and socially cut off, marooned. We had been bombed out of London. After November one had to hoard one's petrol and it was no longer possible to go to London by car. Travel by train became more and more tedious.

All this meant that for the first time in our lives Virginia and I felt we were country dwellers, villagers. And also for the first time we became completely servantless in the Victorian sense. In London before the war we had reduced our establishment to one, a cook, the strange, silent, melancholy Mabel. When the bombing of London began she had come down to us at Rodmell, but, though a typical country woman from the West of England, she disliked the country and hated being away from London. After a few weeks of Rodmell, she could stand it no longer and decided that she preferred the bombs of London. She left us for good to live with her sister and work in a canteen. To be thus finally without servants, without any responsibility for anyone beside ourselves, gave us an additional feeling of freedom and of the dead calm in the centre of the hurricane.

The calm came partly from the routine which established itself, the pleasant monotony of living. We worked all the morning; got our lunch; walked or gardened in the afternoon; played a game of bowls; cooked our dinner; read our books and listened to music; and so to bed. Virginia's diary shows clearly that this life gave her tranquillity and happiness. On October 12 she wrote:

How free, how peaceful we are. No one coming. No servants. Dine when we like. Living near to the bone. I think we've mastered life pretty competently.

And two days later the following long entry gives the depth of her mood and its background (I published it in *A Writer's Diary*, but I quote it here because it is so relevant):

I would like to pack my day rather fuller: most reading must be munching. If it were not treasonable to say so, a day like this is almost too—I won't say happy: but amenable. The tune varies, from one nice melody to another. All is played (today) in such a theatre. Hills and fields; I can't stop looking; October blooms; brown plough; and the fading and freshening of the marsh. Now the mist comes up. And one thing's "pleasant" after another: breakfast, writing, walking, tea, bowls, reading, sweets, bed. A letter from Rose about her day. I let it almost break mine. Mine recovers. The globe rounds again. Behind it—oh yes. But I was thinking I must intensify. Partly Rose. Partly I'm terrified of passive acquiescence. I live in intensity. In London, now, or two years ago, I'd be owling through the streets. More pack and thrill than here. So I must supply that—how? I think book inventing. And there's always the chance of a rough wave: no, I won't once more turn my magnifying glass on that. Scraps of memoirs come so coolingly to my mind. Wound up by those three little articles (one sent today) I unwound a page about Thoby. Fish forgotten. I must invent a dinner. But it's all so heavenly free and easy—L. and I alone. We raised Louie's wages to 15/0 from 12/0 this week.* She is as rosy and round as a

* Louie, since 1932, lived in one of two cottages owned by me in Rodmell. She "did" for us, coming at eight and washing up, making the beds, and cleaning the house—she still does in 1969.

small boy tipped. I've my rug on hand too. Another pleasure. And all the clothes drudgery, Sybil drudgery, society drudgery obliterated. But I want to look back on these war years as years of positive something or other. L. gathering apples. Sally barks. I imagine a village invasion. Queer the contraction of life to the village radius. Wood bought enough to stock many winters. All our friends are isolated over winter fires. Letters from Angelica, Bunny etc. No cars. No petrol. Trains uncertain. And we on our lovely free autumn island. But I will read Dante, and for my trip through English literature book. I was glad to see the *C.R.* all spotted with readers at the Free Library to which I think of belonging.

Another entry in her diary (October 2), which I reprinted in *A Writer's Diary*,* gives vividly her mood that autumn, a kind of quietism and open-eyed contemplation of death. Death was no longer, as it is for all of us all our lives, the end of life, seen always a long way off, unreal, through the wrong end of the telescope of life, but now it was something immediate, extraordinarily near and real, hanging perpetually just above our heads, something which might at any moment come falling with a great bang out of the sky—and annihilate us. "Last night," she wrote, "a great heavy plunge of bomb under the window. So near we both started." Her immediate reaction was: "I said to L.: I don't want to die yet." And then there follows an extraordinarily vivid description of what it would feel like to be killed by a bomb:

> Oh I try to imagine how one's killed by a bomb. I've got it fairly vivid—the sensation: but can't see anything

* Page 340.

but suffocating nonentity following after. I shall think
—oh I wanted another 10 years—not this—and shan't,
for once, be able to describe it.

Death, I think, was always very near the surface of
Virginia's mind, the contemplation of death. It was part
of the deep imbalance of her mind. She was "half in love
with easeful Death". I can understand this, but only intellectually; emotionally it is completely alien to me. Until
I began to grow old, I hardly ever even thought of death.
I knew that it is the inevitable end, but fundamentally I
am a complete fatalist. It is in part perhaps due to the
Jewish tradition, the sceptical fatalism that undermines
even Jehovah in *Ecclesiastes* and deep down in *Job*; and
later in nearly two thousand years of persecution and the
ghettoes of Europe the Jews have learnt that it is a full-
time job to fight or evade life's avoidable evils, the wise
man does not worry about the inevitable. I would accept
the risk of immortality, if I were offered it, but I do not
worry about my inevitable death. As one grows old, one
is forced to think of it, for it grows nearer and nearer; the
time comes when you see that people are surprised to see
that you are still alive, when you know that, if you plant a
tree in your garden, you will not be alive to stand beneath
its branches, or, if you buy a bottle of claret "for laying
down", you will probably die before it has matured.
I have reached this stage when "I shan't be there to see
it" is not academic, for one knows it is the day after
tomorrow. Horace's *pallida mors*, pale death, is sitting
on the horseman's shoulder. But I do not think that
I am boasting or deceiving myself when I say that,
though I resent imminent death, I do not worry about it.

It is fate, the inevitable, and there is nothing to do about it.

Virginia's attitude to death was very different. It was always present to her. The fact that she had twice tried to commit suicide—and had almost succeeded—and the knowledge that that terrible desperation of depression might at any moment overwhelm her mind again meant that death was never far from her thoughts. She feared it and yet, as I said, she was "half in love with easeful Death". Yet in those last months of 1940 with death all round her, when the crash of the falling bomb was quite near us, "I said to L: 'I don't want to die yet'." The reason was that she was calmer and happier than usual. This was largely due to the ease and contentment of her writing. It is strange and ironical that *Between the Acts*, which was so soon to play a great part in her breakdown and suicide, caused her so little trouble and worry in the actual finishing of it. "Never had a better writing season. P.H." (i.e. *Between the Acts*) "in fact pleases me", she wrote in her diary on October 6, 1940. On November 5, "I am very 'happy' as the saying is: and excited by *P. H.*" And when she finished the book on November 23, she wrote:

> I am a little triumphant about the book. I think it's an interesting attempt in a new method. I think it's more quintessential than the others. More milk skimmed off. A richer pat, certainly a fresher than that misery *The Years*. I've enjoyed writing almost every page.

It is significant that, on the morning when she finished *Between the Acts*, she was already thinking of the first chapter of her next book. As was always the case with her, before she had finished a book she already had in her mind some outline of the theme and form of another. The book which was to follow *Between the Acts*, and which she did not live to write, was to be called *Anon*, a "fact supported book"; "I think," she wrote, "of taking my mountain top—that persistent vision—as a starting point". So the last months of 1940 passed away for Virginia in a real—and yet false—tranquillity. One amusing incident, characteristic of Virginia's immaculate feminism, which—particularly with regard to *Three Guineas*—has been castigated by many male critics, but which I personally feel to have been eminently right, happened in November. Morgan Forster asked her whether he might propose her for the London Library Committee. But years ago Morgan himself in the London Library itself, meeting Virginia and talking about its organization or administration, had "sniffed about women on the Committee". Virginia at the time made no comment, but she said to herself: "One of these days I shall refuse." So now on Thursday, November 7, 1940, she had some quiet satisfaction in saying No. "I don't want to be a sop—face saver," she wrote in her diary. Many people think this kind of thing petty, trivial. I don't agree: I think something of profound social importance is uncovered in this obvious triviality. One of the greatest of social evils has always been class subjection and class domination. The struggle to end the subjection of women has been bitter and prolonged; it was not by any means over in 1940, nor in 1968 either. The male monopoly and vested interest can be seen of course in a small way—in the all male Committees of institutions like the London Library. In 1940 Virginia had already been a member of the Library for about forty years. She was eminently fitted to be a member of the Committee and so were many other women who were members of the Library. There is no doubt that, if they had been men, many of them would have been elected to the Committee. It was fantastic that before 1940 none of them had been elected and that Morgan could sniff at the idea of a woman on the Committee. And the spectacle of the male committee suddenly anxious to elect one woman—"a sop, a face saver"—in order to show their sexual open-mindedness is by no means uncommon. The egalitarian sees in these trivialities a real social significance.*

* An interesting example of the complacent male monopoly occurred twenty-seven years after this in the Royal Horticultural Society. The administrative organ of this large, affluent society is a large Council. The Council is elected by the members at an annual general meeting, but, as everyone with experience of annual general meetings and committees knows, the committees or councils of institutions like the London Library and the Royal Horticultural Society are nearly always self-reproducing. The ordinary members are quiescent and acquiescent and do not propose people for election; the committee, if there are three vacancies, proposes exactly three names for election and they are automatically elected. In the year 1967 the Council of the R.H.S. was entirely male, and, when Lady Enid Jones and some others wrote to *The Times* protesting against the absence of female Councillors, the President, Lord Aberconway, defending the Establishment, seemed to imply that, all through the many years of the society's existence, the officers and Council had searched for a female horticulturist worthy to sit with them on the Council—and had failed to find one. This is surprising when one remembers the many famous names of women horticulturists, e.g. Miss Jekyll, Vita Sackville-West, Mrs Earle. Perhaps it was still more surprising that within a few months Lord Aberconway and the Council succeeded in unearthing a woman fit to serve. Lord Aberconway, at the annual general meeting, went out of his way to compliment the society and Mrs Perry on the fact that she would be

In Virginia's diary for the last two months of 1940 there is evidence of her tranquillity. There is, however, one exasperated and slightly unbalanced entry which might be thought to contradict this, but in fact Virginia all her life at any moment might have suffered this kind of short and sharp spasm of exasperation. Only hindsight could read something abnormally serious into this outburst. It is however a curious outburst. On November 28 I gave a lecture (I do not remember on what) to the Workers' Educational Association. Like everyone else, we were often asked—Virginia would have said pestered—to do this kind of thing. On November 29 Virginia wrote in her diary:

> Many many deep thoughts have visited me. And fled. The pen puts salt on their tails; they see the shadow and fly. I was thinking about vampires. Leeches. Anyone with 500 a year and education is at once sucked by the leeches. Put L and me into Rodmell pool and we are sucked—sucked—sucked. I see the reason for those who suck guineas. But life—ideas—that's a bit thick. We've exchanged the clever for the simple. The simple envy us our life. Last night L's lecture attracted suckers.

It was only in the first days of 1941 that the deep disturbance in her mind began to show itself clearly. I shall continue to quote from her diary because her own words are

the first "lady member"; he added: "It is my further personal hope ... that before long we may decide that a second lady may be the most suitable candidate to fill a vacancy on the Council." The patronizing, complacent dictatorship of the male horticulturist in 1968 seems to me some justification of Virginia's contemptuous irritation with Morgan and the London Library in 1940.

more revealing and authentic than my memory. The entry for January 9 is again strange, showing her preoccupation with death:

A blank. All frost. Still frost. Burning white. Burning blue. The elms red. I did not mean to describe, once more, the downs in snow; but it came. And I can't help even now turning to look at Asheham down, red, purple, dove blue grey, with the cross* so melodramatically against it. What is the phrase I always remember—or forget. Look your last on all things lovely. Yesterday Mrs. Dedman was buried upsidedown. A mishap. Such a heavy woman, as Louie put it, feasting spontaneously upon the grave. Today she buries the Aunt whose husband saw the vision at Seaford. Their home was bombed by the bomb we heard early one morning last week. And L. is lecturing and arranging the room. Are these the things that are interesting? that recall: that say Stop, you are so fair? Well, all life is so fair, at my age. I mean, without much more of it I suppose to follow. And t'other side of the hill there'll be no rosy blue red snow.

Then round about January 25, I think, the first symptoms of serious mental disturbance began to show themselves. She fell into what she called a "trough of despair". It was a sudden attack and it lasted ten or twelve days. There was something strange about it, for, when it passed off, she said herself that she could not remember why she had been depressed. It did not appear to be connected with her revising *Between the Acts*—indeed, on February 7 she

* The stone cross on the Rodmell church is visible from the window of our sitting-room silhouetted against the down.

noted that she had been writing with some glow. Nevertheless, I am sure that what was about to happen was connected with the strain of revising the book and the black cloud which always gathered and spread over her mind whenever, a book finished, she had to face the shock of severing as it were the mental umbilical cord and send it to the printer—and finally to the reviewers and the public. I did not at first realize quite how serious these symptoms were, though I at once became uneasy and took steps which I will describe later. One thing which deceived me was the suddenness of this attack. For years I had been accustomed to watch for signs of danger in Virginia's mind; and the warning symptoms had come on slowly and unmistakably; the headache, the sleeplessness, the inability to concentrate. We had learnt that a breakdown could always be avoided, if she immediately retired into a hibernation or cocoon of quiescence when the symptoms showed themselves. But this time there were no warning symptoms of this kind. The depression struck her like a sudden blow. Looking back over what happened I can now see that once before there had been an even more sudden mental disturbance, a sudden transition from mental stability to disorder. In that case it was even more catastrophic. I described what happened in *Beginning Again**. It happened early in 1915 when we were staying in lodgings on Richmond Green. Virginia seemed to have recovered from the terrible breakdown which had lasted for the better part of a year. One morning she was having breakfast in bed and I was sitting by the bedside talking with her. She was calm, well, perfectly sane. Suddenly she became violently excited, thought her mother was in

* Page 172.

the room, and began talking to her. That was the beginning of the long second stage in a complete mental breakdown.

I think it must have been about the middle of January that I began to be uneasy about Virginia and consulted Octavia Wilberforce. Octavia was a remarkable character. Her ancestors were the famous Wilberforces of the anti-slavery movement; their portraits hung on her walls and she had inherited their beautiful furniture and their fine library of eighteenth-century books. Her family was closely connected with Virginia's, both having their roots in the Clapham Sect. Octavia had been born and bred in a large house in Sussex, a young lady in a typical country gentleman's country house. But though she was always very much an English lady of the upper middle class, she was never a typical young lady. *Illi robur et aes triplex circum pectus erat*—oak and triple brass were around her breast—in all the important things of life. She was large, strong, solid, slow growing, completely reliable, like an English oak. Her roots were in English history and the English soil of Sussex, and, in her reserved way, she was deeply attached to both. She was already a young lady when she decided that she must become a doctor. It was a strange, disquieting decision, for in Sussex country houses in those days young ladies did not become doctors; they played tennis and went to dances in order to marry and breed more young ladies in more country houses who would breed still more young ladies in still more country houses. Octavia's idea was not thought to be a good one by her family, and she received no encouragement there. Another difficulty was that her education as a young lady was not the kind which made it easy for her to pass the necessary examinations to qualify as a doctor. But her quiet determination, the oak and triple brass enabled her to overcome all difficulties. She became a first-class doctor in Brighton.

Octavia practised as a doctor in Montpelier Crescent, Brighton, and lived there with Elizabeth Robins. It was in 1928 that we got to know them. Virginia had been awarded the Femina Vie Heureuse prize, and one afternoon in May we went to the French Institute in Cromwell Road for the ceremony of the prize giving. After Hugh Walpole's speech and the presentation of the prize, the usual brouhaha began, and "little Miss Robins, like a redbreast, creeping out", introduced herself to Virginia. She had known Leslie Stephen and the whole Stephen family when Vanessa and Virginia were small children. She had the gift of vivid visual memory and of vividly describing what she saw down the wrong end of the telescope of memory long ago and far away. She described Virginia's mother so that one saw her for the first time a living woman, a very different figure from the saintly dying duck of her husband's memories and even of Mrs Cameron's photographs: "she would suddenly say something so unexpected, from that Madonna face, one thought it *vicious*". We asked Elizabeth to dinner with us in London and later went to see her in Brighton—and so we got to know Octavia.

Elizabeth was an even more remarkable woman than Octavia. She was born in Kentucky in 1862, a young lady belonging to the old slave-owning American aristocracy of the South. She did in Kentucky what Octavia was to do later on in Sussex; with extraordinary strength of mind and determination she broke the fetters of family and class, the iron laws which prescribe the life and behaviour of young ladies whether they be the Greek Antigone 600 years before Christ in Thebes or 2,500 years later Elizabeth in Kentucky, U.S.A., and Octavia in Lavington, Sussex. Elizabeth decided that she must become an actress, an unheard-of thing for a young lady in a Southern State, daughter of a banker and grand-daughter of a very formidable grandmother. The family was adamant and her father whisked her away to the Rocky Mountains in the hope that there she would forget all about the stage. But her will was as rocky as the mountains and her family had to give way. She went on the stage, toured America, came to London, and became a famous actress. She was the first actress to play the parts of the great heroines of Ibsen's plays, Hedda Gabler, Hilda in *The Master Builder* and Nora in *A Doll's House*. I think that she must have been a great actress. She was sixty-six years old when I first knew her and she was ninety when she died. But even when she was a very old woman, there was something magnetic in her when she spoke about the great characters of Shakespeare or Ibsen, and I felt the same passionate dedication to her art that is so noticeable in another great actress, Peggy Ashcroft, and in the great Russian ballerina, Lydia Lopokova. Her talents or genius, however, were not confined to the stage. At the height of her career she threw up everything and set off by herself to the frozen wilderness of Alaska in search of her beloved brother Raymond. He was a strange, gifted, wayward character; he had the same mercurial vitality that she had, and he also suffered from the curious psychological kink, which I have known in two other men, occasionally all through his life an irresistible impulse would come upon him to run away from his life, to disappear. On this occasion he joined the gold rush to the mining camps in Klondike—and after that complete silence.

Elizabeth, as I said, set off entirely by herself to find him. By that time she had become not only a distinguished actress but a well-known figure in the literary society of London. For such a young woman to start off entirely by herself for a mining-camp almost in the Arctic Circle was a dreadful, impossible thing to do in 1900. Her friends were horrified, but Elizabeth was fearless and indomitable. She disappeared into the snows of Alaska, but she found Raymond and stayed with him in the wild mining-town until she got ill and had to return to London. She did not return to the stage, but wrote a best-selling novel with an Arctic Alaskan background, *The Magnetic North*. She was a gifted writer, a fairly prolific novelist; she also wrote two vivid autobiographical works, *Raymond and I*, which described her odyssey in search of her brother, and *Both Sides of the Curtain*, which told of her life on the stage and of the many distinguished men and women whom she had known.

It was in 1908 that the friendship between Octavia and Elizabeth began. When we first got to know them, they were living, as I said, in Brighton. Elizabeth also owned a lovely Sussex farmhouse surrounded by a good deal of grazing land in Henfield. She had lived there until 1927 when she turned it into a Home of Rest for overworked professional women, a charitable trust over which Octavia presided medically and administratively. I give these details and those that follow because of the part which Octavia and Elizabeth played in the last months of Virginia's life and because afterwards my friendship with them led to my being concerned with their affairs and their relationship and also with the Backsettown Trust.

Octavia's relation to Elizabeth was that of a devoted daughter. If you had searched the earth from Kentucky in the United States to Lavington in Sussex, you would never and nowhere have found two other women more different from each other than they were. Elizabeth was, I think, devoted to Octavia, but she was also devoted to Elizabeth Robins; when we first knew her, she was already an elderly woman and a dedicated egoist, but she was still a fascinating as well as an exasperating egoist. When young she must have been beautiful, very vivacious, a gleam of genius with that indescribably female charm which made her invincible to all men and most women. One felt all this still lingering in her as one sometimes feels the beauty of summer still lingering in an autumn garden. She was not an easy companion, for she had that vampire nature which some old people develop which enables them to drain the strength and vitality of the young so that the older they grow the more invincible, indefatigable, imperishable they become. After the war, when she returned from Florida to Brighton, a very old and frail woman, she used every so often to ask me to come and see her and give her advice on some problem. I would find her in bed, surrounded by boxes full of letters, cuttings, memoranda, and snippets of every sort and kind. In stamina I am myself inclined to be invincible, indefatigable, and imperishable, and I was nearly twenty years younger than Elizabeth, but after two or three hours' conversation with her in Montpelier Crescent, I have often staggered out of the house shaky, drained, and debilitated as if I had just recovered from a severe attack of influenza.

She was so much an individual and so complex that it is impossible to paint a complete and satisfactory portrait of her, but it is worth recording one strange habit of hers. I do not think that throughout her life of ninety years she can ever have destroyed a single letter, document, or scrap of paper which concerned her or even merely passed through her hands. In her will she named Octavia and me as her executors, and we found literally mountains of letters and documents in the house in Montpelier Crescent and in dozens of trunks stored in a furniture depository. I will give one example of her squirrel or jackdaw hoarding habit. When the bombing became severe in the war, her brother Raymond induced her—much against her will—to come to the U.S.A. and stay in Florida. One day in Florida she bought a dozen bottles of soda water from a store; her letter ordering them and the receipt for what she paid for them were filed with all the other myriad documents of the utmost importance and of no importance at all. Later on she returned the empty bottles and ordered another dozen; the second dozen was sent together with an account which did not allow for the returned dozen. She wrote a letter pointing out the omission and a corrected account was sent to her and paid by her. All these documents were filed, docketed, brought to England at the end of the war, and stored in a Brighton depository.

I feel that what I have written in the previous paragraphs will give to those who never saw or heard her a one-sided and inaccurately unfavourable picture of Elizabeth. It gives no idea of the charm and lovableness which age and egoism had not destroyed. Virginia fascinated her and I think she was fond of both of us. In the summers before the war, when we were in Rodmell for August and September, Octavia would from time to time bring Elizabeth to see us, and we would sit under the chestnut-tree in the garden talking. Elizabeth was an incorrigible talker. She had the gift, which is not uncommon among Irish women and women from the Southern States of America, of telling old tales of their youth poetically, romantically. Elizabeth had a beautiful voice; as an actress she could both be and act herself; the child in the Deep South, in the long white house, the languorous heat, the soft opulent life, the all-enveloping family with the adored, dominating, devouring grandmother. Virginia and I were entranced by this saga.

In the summer of 1939 and the first half of 1940 we saw Elizabeth and Octavia in this way from time to time and sometimes went to see them in Brighton. In the latter part of 1940 Elizabeth was induced to go to America, but we still saw Octavia. She had, to all intents and purposes, become Virginia's doctor, and so the moment I became uneasy about Virginia's psychological health in the beginning of 1941 I told Octavia and consulted her professionally. The desperate difficulty which always presented itself when Virginia began to be threatened with a break-down—a difficulty which occurs, I think, again and again in mental illnesses—was to decide how far it was safe to go in urging her to take steps—drastic steps—to ward off the attack. Drastic steps meant going to bed, complete rest, plenty of food and milk. But part of the disease was to deny the disease and to refuse the cure. There was always the danger of reaching the point when, if one continued to urge her to take the necessary steps, one would only increase not only the resistance but her terrible depression. Food in any case was a problem owing to rationing and shortages, and Octavia, who ran a farm at Henfield with a herd of Jersey cows, in January and February used to come to tea with us once a week bringing with her milk and cream.

Twelve days after Virginia's "trough of depression" the mood had passed away and she wrote in her diary: "Why was I depressed? I cannot remember". That was on February 7, and on February 11 we went to Cambridge for two nights and visited The Hogarth Press at Letchworth. Virginia seemed to enjoy this—the usual round of a visit to Cambridge, seeing Pernel Strachey, Principal of Newnham, and dining with Dadie Rylands in King's. Then we had a round of visitors: Elizabeth Bowen for two nights, Vita Sackville-West, and Enid Jones. Again Virginia seemed to enjoy a good deal of this, and I was less uneasy. Something of the state of her mind may, perhaps, be shown by the fact that on February 26 she recorded the following in her diary:

Yesterday in the ladies' lavatory at the Sussex Grill in Brighton I heard: "She's a little simpering thing. I don't like her. But then he never did care for big women. (So to Bert.) His eyes are so blue. Like blue pools. So's Gert's. They have the same eyes. Only her teeth part a little. He has wonderful white teeth. He always had. It's fun having the boys. . . . If he don't look out he'll be court martialled."

They were powdering and painting, these common little tarts, while I sat behind a thin door, p—ing as quietly as I could.

Then at Fuller's. A fat smart woman in red hunting cap, pearls, check skirt, consuming rich cakes. Her shabby dependant also stuffing. Hudson's van unloading biscuits opposite. The fat woman had a louche large white muffin face. T'other was slightly grilled. They ate and ate. Talked about Mary. But if she's ill, you'll have to go to her. You're the only one. . . . But why should she be? . . . I opened the marmalade but John doesn't like it. And we have two pounds of biscuits in the tin upstairs. . . . Something scented, shoddy, parasitic about them. Then they totted up cakes. And passed the time o' day with the waitress. Where does the money come from to feed these fat white slugs? Brighton a love corner for slugs. The powdered, the pampered, the mildly improper. I invested them in a large house in Sussex Square. We cycled. Irritated as usual by the blasphemy of Peacehaven. Helen has fallen through, I mean the house I got her with X, the day X lunched here with Vita: and I felt so untidy yet cool; and she edgy and brittle. No walks for ever so long. People daily. And rather a churn in my mind. And some blank spaces. Food becomes an obsession. I grudge giving away a spice bun. Curious-age, or the war? Never mind. Adventure. Make solid. But shall I ever write again one of those sentences that give me intense pleasure? There is no echo in Rodmell—only waste air. . . . I spent the afternoon at the School, marbling paper. Mrs D discontented and said: There's no life in these children, comparing them with Londoners, thus repeating my own comment after that long languid meeting at Chavasses. No life: and so they cling to us. This is my conclusion. We pay the penalty for our rung in society by infernal boredom.

There are ominous signs in this entry. She had just finally finished *Between the Acts* and had given it to me to read. I saw at once now the ominous symptoms and became again very uneasy. After the entry in her diary on February 26, quoted above, there are only two entries before she committed suicide on March 28, one on March 8 of which I printed part in *A Writer's Diary* and the last on March 24. I now give the unpublished portion of the March 8 entry and the final entry of March 24, since they show, I think, very clearly the state of her mind in those last days:

Sunday, March 8th
. . . Last night I analysed to L. my London Library complex. That sudden terror has vanished; now I'm plucked at by the H. Hamilton lunch that I refused. To right the balance, I wrote to Stephen and Tom: and I will write to Ethel and invite myself to stay; and then to Miss Sharp who presented me with a bunch of violets. This to make up for the sight of Oxford Street and Piccadilly, which haunts me. Oh dear yes, I shall conquer this mood. It's a question of being open sleepy, wide eyed at present: letting things come one after another. Now to cook the haddock.

March 24th
She had a nose like the Duke of Wellington and great horse teeth and cold prominent eyes. When we came in she was sitting perched on a three-cornered chair with knitting in her hands. An arrow fastened her

collar. And before five minutes had passed she had told us that two of her sons had been killed in the war. This one felt was to her credit. She taught dressmaking. Everything in the room was red brown and glossy. Sitting there I tried to coin a few compliments. But they perished in the [?] sea between us. And then there was nothing.

A curious seaside feeling in the air today. It reminds me of lodgings in a parade at Easter. Everyone leaning against the wind, nipped and silenced. All pulp removed.

This windy corner and Nessa is at Brighton and I am imagining how it would be if we could infuse souls.

Octavia's story. Could I englobe it somehow? English youth in 1900.

Two long letters from Shena and O. I can't tackle them, yet enjoy having them. L. is doing the rhododendrons.

Shena was Lady Simon of Wythenshawe. Octavia's story refers to a vague scheme of Virginia's. Whenever Octavia came to see us, Virginia tried to get her to "tell the story of her life" and she had this vague idea of perhaps making it into a book. It is clear therefore from this entry that even four days before her suicide she could be thinking of writing another book. On the other hand there are signs of deep disturbance in these last entries. There is a note in my diary on March 18 that she was not well and in the next week I became more and more alarmed. I am not sure whether early in that week she did not unsuccessfully try to commit suicide. She went for a walk in the water-meadows in pouring rain and I went, as I often did, to meet her. She came back across the meadows soaking wet, looking ill and shaken. She said that she had slipped and fallen into one of the dykes. At the time I did not definitely suspect anything, though I had an automatic feeling of desperate uneasiness. On Friday, March 21, Octavia came to tea and I told her that I thought Virginia on the verge of danger. On Monday, March 24, she was slightly better, but two days later I knew that the situation was very dangerous. Desperate depression had settled upon Virginia; her thoughts raced beyond her control; she was terrified of madness. One knew that at any moment she might kill herself. The only chance for her was to give in and admit that she was ill, but this she would not do. Octavia had been coming to see us about once a week, bringing cream and milk. These visits were, so far as Virginia was concerned, just friendly visits, but I had told Octavia how serious I thought Virginia's condition was becoming and from our point of view, the visits were partly medical. On Wednesday, March 26, I became convinced that Virginia's mental condition was more serious than it had ever been since those terrible days in August 1913 which led to her complete breakdown and attempt to kill herself. The terrifying decision which I had to take then once more faced me. It was essential for her to resign herself to illness and the drastic regime which alone could stave off insanity. But she was on the brink of despair, insanity, and suicide. I had to urge her to face the verge of disaster in order to get her to accept the misery of the only method of avoiding it, and I knew at the same time that a wrong word, a mere hint of pressure, even a statement of the truth might be enough to drive her over the verge into suicide. The memory of

1913 when the attempted suicide was the immediate result of the interview with Dr Head haunted me.*

Yet one had to take a decision and abide by it, knowing the risk—and whatever one decided, the risk was appalling. I suggested to Virginia that she should go and see Octavia and consult her as a doctor as well as a friend. She agreed to this and next day I drove her to Brighton. She had a long talk with Octavia by herself and then Octavia came into the front room in Montpelier Crescent and she and I discussed what we should do. We stood talking by the window and suddenly just above the roofs of the houses a German bomber flew, almost as it were just above our heads, following the line of the street; it roared away towards the sea and almost immediately there was a crash of exploding bombs. We were so overwhelmed by our problem and so deep in thought and conversation that the sight and sound were not at the moment even consciously registered, and it was only some time after I had left Brighton and was driving back to Lewes that I suddenly remembered the vision of the great plane just above our heads and the crash of the bombs.

It seemed possible that Octavia's talk had had some effect upon Virginia and it was left that she would come and see Virginia again in Rodmell in a day or two. We felt that it was not safe to do anything more at the moment. And it was the moment at which the risk had to be taken, for if one did not force the issue—which would have meant perpetual surveillance of trained nurses—one would only have made it impossible and intolerable to her if one attempted the same kind of perpetual surveillance by oneself. The decision was wrong and led to the dis-

* See *Beginning Again*, pp. 150, 151, and 154-7.

aster. The next day, Friday, March 28, I was in the garden and I thought she was in the house. But when at one o'clock I went in to lunch, she was not there. I found the following letter on the sitting-room mantelpiece:*

Dearest,
    I feel certain that I am going mad again. I feel we can't go through another of those terrible times. And

* Later on I found the following letter on the writing block in her work-room. At about eleven on the morning of March 28 I had gone to see her in her writing-room and found her writing on the block. She came into the house with me, leaving the writing block in her room. She must, I think, have written the letter which she left for me on the mantelpiece (and a letter to Vanessa) in the house immediately afterwards.

Dearest,
    I want to tell you that you have given me complete happiness. No one could have done more than you have done. Please believe that. But I know that I shall never get over this: and I am wasting your life. Nothing anyone says can persuade me. You can work, and you will be much better without me. You see I can't write this even, which shows I am right. All I wish to say is that until this disease came on me we were perfectly happy. It was all due to you. No one could have been so good as you have been from the very first day till now. Everyone knows that.

                                                              V.

    You will find Roger's letters to Mauron in the writing-table drawer in the Lodge. Will you destroy all my papers?

The following is the letter which she wrote to Vanessa:

Sunday

Dearest,
    You can't think how I loved your letter. But I feel that I have gone too far this time to come back again. I am certain now that I am going mad again. It is just as it was the first time, I am always hearing voices, and I know I shan't get over it now. All

I shan't recover this time. I begin to hear voices, and I can't concentrate. So I am doing what seems the best thing to do. You have given me the greatest possible happiness. You have been in every way all that anyone could be. I don't think two people could have been happier till this terrible disease came. I can't fight any longer. I know that I am spoiling your life, that without me you could work. And you will I know. You see I can't even write this properly. I can't read. What I want to say is I owe all the happiness of my life to you. You have been entirely patient with me and incredibly good. I want to say that—everybody knows it. If anybody could have saved me it would have been you. Everything has gone from me but the certainty of your goodness. I can't go on spoiling your life any longer.

I don't think two people could have been happier than we have been.

V.

When I could not find her anywhere in the house or garden, I felt sure that she had gone down to the river. I ran across the fields down to the river and almost im-

---

I want to say is that Leonard has been so astonishingly good, every day, always; I can't imagine that anyone could have done more for me than he has. We have been perfectly happy until the last few weeks, when this horror began. Will you assure him of this? I feel he has so much to do that he will go on, better without me, and you will help him.

I can hardly think clearly any more. If I could I would tell you what you and the children have meant to me. I think you know.

I have fought against it, but I can't any longer.

Virginia.

mediately found her walking-stick lying upon the bank.
I searched for some time and then went back to the house
and informed the police. It was three weeks before her
body was found when some children saw it floating in the
river. The horrible business of the identification and inquest took place in the Newhaven mortuary on April 18
and 19. Virginia was cremated in Brighton on Monday,
April 21. I went there by myself. I had once said to her
that, if there was to be music at one's cremation, it ought
to be the cavatina from the B flat quartet, op. 130, of
Beethoven. There is a moment at cremations when the
doors of the crematorium open and the coffin slides slowly
in, and there is a moment in the middle of the cavatina
when for a few bars the music, of incredible beauty, seems
to hesitate with a gentle forward pulsing motion—if
played at that moment it might seem to be gently propel-
ling the dead into eternity of oblivion. Virginia agreed
with me. I had always vaguely thought that the cavatina
might be played at her cremation or mine so that these
bars would synchronize with the opening of the doors and
the music would propel us into eternal oblivion. When I
made the arrangements for Virginia's funeral, I should
have liked to arrange this, but I could not bring myself to
do anything about it. It was partly that, when I went to old
Dean at the top of the village, whom we had known for
nearly a quarter of a century, to get him to make the
arrangements, it seemed impossible to discuss Beethoven's
cavatina with him, and impossible that he could supply
the music. But it was also that the long-drawn-out horror
of the previous weeks had produced in me a kind of inert
anaesthesia. It was as if I had been so battered and beaten
that I was like some hunted animal which exhausted can

only instinctively drag itself into its hole or lair. In fact (to my surprise) at the cremation the music of the "Blessed Spirits" from Gluck's *Orfeo* was played when the doors opened and the coffin disappeared. In the evening I played the cavatina.

I buried Virginia's ashes at the foot of the great elm tree on the bank of the great lawn in the garden, called the Croft, which looks out over the field and the water-meadows. There were two great elms there with boughs interlaced which we always called Leonard and Virginia. In the first week of January 1943, in a great gale one of the elms was blown down.

Chapter Two

THE HOGARTH PRESS

WHEN the war broke out, The Hogarth Press was in a flourishing condition. A year before, in 1938, it had suffered a revolutionary change in its constitution and management. I had taken John Lehmann into partnership. This was effected by Virginia formally selling her fifty per cent interest in the Press to him; John and I then entered into a partnership agreement which gave us equal rights in the business, though he was to undertake the day-to-day management of it. In 1931, at the age of twenty-four, without publishing experience, he had come to us as manager, but as I related in *Downhill All the Way*,* the venture was not a success and he left us in 1932. In the six years between his leaving us as manager and returning to us as a partner I ran the Press on my own in my own way with a woman manager, and I gave up the idea of finding a partner. Both Virginia and I got a great deal of enjoyment out of these six years of publishing on our own in our own peculiar way. If one cares for books and literature, as we did, there is real pleasure in finding good writers and personally publishing their works. We had, I think, a remarkable list in those six years. We began the publication of Rilke's poetry in Leishman's translations: *Poems* in 1934, *Requiem* in 1935, *Sonnets to Orpheus* in 1936, and *Later Poems* in 1938. The following are some of the other books published by us during the period:

* Pages 172-6.

Freud's *An Autobiographical Study* (1935) and *Inhibitions, Symptoms and Anxiety* (1936); Isherwood's *Mr Norris Changes Trains* (1935) and *Lions and Shadows* (1938); Ivan Bunin's *Grammar of Love* (1935); Laurens van der Post's *In a Province* (1934); Bertrand Russell's *The Amberley Papers* (1937); Vita Sackville-West's *Pepita* (1937); Virginia's *The Years* (1936) and *Three Guineas* (1938).

From my point of view the business was financially very successful. We never considered it in any way as a means of making a living. I had always treated it as a half-time, or, more strictly, quarter-time occupation and we deliberately fought against its expansion into a larger scale business. We were determined to publish only books which we thought worth publishing and our aim was to limit our list to a maximum of round about 20 new books a year. This was by no means easy. We were often offered books which, as the saying is, "any publisher would like to have on his list", but which we refused because they would have swelled our list more than we wanted to see it swell. Our interest in the business of writing and publishing books also led us into continually having ideas for new books or series, and, when our enthusiasm induced us to get the books written, this from time to time made our list longer than we liked. For instance, a year before John came to us I started a series which I called World-Makers and World-Shakers. It was a series of short biographies for young people which would attempt to explain history to them through the lives of great men and women, and at the same time present history from a modern and enlightened point of view. I had hoped to get the books used in schools. The hope was not fulfilled, and, though we sold out the edition of the four books which we published, we never got the sale we wanted and did not go on with the venture. This was partly due to the fact that we were so soon overwhelmed by the war and the difficulty of getting paper. But I still think that the idea was a good one, and our first four books were extremely interesting. They were: *Socrates* by Naomi Mitchison and R. H. S. Crossman; *Joan of Arc* by V. Sackville-West; *Mazzini, Garibaldi and Cavour* by Marjorie Strachey; and *Darwin* by L. B. Pekin.

I said above that The Hogarth Press was by 1935 a successful business financially. In the three years before John came into it my income from it was annually over £1,000. For a quarter-time occupation this was satisfactory, but in fact Virginia and I did not need (or want) to make £1,000 a year by publishing. We made enough from writing to live the kind of life we wanted to live without bothering about money. We knew the kind of life we wanted to live and we would not have altered it however much money we might make. We were much better off in 1935 than we had been in 1925, but in fact fundamentally we had not altered our way of life. It is extremely pleasant to have plenty of money, particularly for anyone who has previously had little or none. There are two reasons why it is pleasant: first, if you have plenty of money you need no longer think about money; secondly, you can make yourself physically comfortable. We found that it works in mysterious, and often unexpected, ways. For instance, the Victorian domestic system, in which we were both brought up, assumed that one's comfort depended upon having servants living in the house, and we still had a cook and house parlourmaid in the nineteen-twenties. But it is really much more comfortable not to have servants living in, provided that you have enough money to organize your physical life without them. It was only when we had become comfortably well off that we dispensed with the comfort of a cook. We bought the kinds of things which make it easy "to do for yourself". And we bought two cottages in the village. In one we put a gardener, Percy, and in the other a young married woman, Louie Everest.

Percy and Louie were both remarkable people, descendants of a long line of agricultural workers and with their roots deep in Sussex. Percy, who is now dead, lived in my cottage and cultivated my garden for twenty-five years; Louie is still living in my cottage and working for me after thirty-six years. I liked Percy very much, though he was the most pigheaded man I have ever known. When he had been with me for over twenty years he got cataract in both eyes and some years later went into hospital for an operation. If he had had the operation, it would almost certainly have been successful, but for some unexplained reason two days before the operation was to take place, in the middle of the night Percy got very angry, said that he would not have the operation at all, and must leave the hospital immediately. He made the nurses summon his wife; she arrived early in the morning in a taxi and took him home. His sight became worse and worse and in the last years of his life he was completely blind. He might have come out of a novel by Balzac or perhaps Zola or a tale of Maupassant—very, very English, a character not uncommon in rural England from the time of Shakespeare. It is strange that, though he was so English, I think of three French novelists in whose pages one might have met him, but no English writer. One only has to write that, and of course immediately Hardy and his gallery of yokels rise up before one. But Percy and many of the other Sussex agricultural workers whom I have known could not have come out of the Wessex novels. There was an element of grim, granitic tragedy not very far below the surface in them which is very different from the fatalistic tragedy indigenous to the softer and gentler Dorsetshire. They would, as I said, have found themselves more at home in Brittany and Normandy with the peasants of Zola or Maupassant.

Percy's wife, who came from a different class from his, her father having owned a milling business in East Anglia, inherited over £10,000 from an aunt. Having become well-to-do rentiers, they hardly altered their way of life, for they continued to live in my cottage and Percy went on working as my gardener until he went blind. They then bought a house in Lewes.

Louie, as I said, is a no less remarkable character than was Percy. Her native intelligence is extraordinary and she has that rare impersonal curiosity which the Greeks recognized as the basis of philosophy and wisdom. I, as her employer, have known her in daily life for thirty-six years, and, though she is shrewd, critical, and sceptical, I have never heard a complaint from her, and she is, I think, the only person whom I have ever known to be uniformly cheerful and with reason for her cheerfulness.

It was due to Percy and Louie, as I said, that we were able to live in comfort without servants when we were in Rodmell, and it was only when we became more or less affluent that we could afford Percy and Louie. The annual income of £1,000 from The Hogarth Press was part of the affluence, but by 1938 we did not need it. The day-to-day running of the Press had, after three years of it, become a burden. In those days I edited the *Political Quarterly*, was a member of the Civil Service Arbitration Tribunal, was secretary of two Labour Party Advisory Committees, and did a good deal of work in the Fabian Society. I did a certain amount of reviewing and every now and again acted as editor of the *New Statesman* when Kingsley Martin wanted to go off abroad. What I most wanted to do was to write books, but I found it difficult to get the time to do this. Though I reckoned the Press to be for me a quarter-time occupation, it was the most exacting and insistent of all my activities. It tied us to the basement in Tavistock Square in a way which was irksome to both of us, for it meant that, as is always the case with a one-man business, it made it very difficult to get completely away from it for any length of time.

By 1938, therefore, we had slipped back into the rather absurd position in which we had wobbled from 1923 to 1932. Should we give the whole thing up or should we try once more to get someone to come in with us and take on the day-to-day management of the business? When John Lehmann reappeared out of the blue, or rather out of the continent of Europe, and called upon us, it became clear that, despite his abrupt leaving of the Press in 1932, bygones had become bygones and he was anxious to return to it. In his two volumes of autobiography he has given his own account of his return, of the seven years of partnership with me, and of his second abrupt withdrawal in 1945. Everyone's vision of the past is more or less distorted by his own personal emotions and prejudices, and my recollection of what happened to the Press in those

The *New Statesman* Board of Directors
Left to right: Kingsley Martin, Leonard Woolf, V. S. Pritchett,
John Freeman, Gerald Barry, John Morgan, Jock Campbell

The author after receiving an honorary degree at Sussex University

troubled years from 1938 to 1945 differs, not unnaturally, in some respects from John's. He says that we first offered to sell the Press outright to him, but that he could not raise the money. We were always talking of giving the whole thing up, but I do not think that we ever seriously thought of selling it outright to him in 1938. It was always a question of a partnership and we had no difficulty in agreeing on its terms.

If we brought to John The Hogarth Press, with all that it had published in its twenty-one years of existence, John brought to us and to the Press *New Writing*. *New Writing* was, justifiably, the apple of John's eye, his publishing ewe lamb. It was a good lamb, of which he was justifiably very proud. In 1932, when he was with us as manager, we published in the Hogarth Living Poets Series, Vol. 24, a slim volume edited by Michael Roberts, *New Signatures*, and we followed this in 1933, the year after he left us, by publishing *New Country*, an anthology of "Prose and Poetry by the Authors of *New Signatures*", also edited by Michael Roberts. *New Signatures*, the conception of which was due to Roberts and John, was a landmark in modern poetry. It contained the work of nine poets. We had already published something of six of them before 1931: C. Day Lewis, Julian Bell, Empson, Eberhart, William Plomer, and John Lehmann himself principally in two volumes of Cambridge Poetry. Roberts and John brought in three new poets: Auden, Stephen Spender, and A. S. J. Tessimond. In *New Country* Michael Roberts threw a still wider net, for, in addition to the original nine, he included among others Christopher Isherwood (whose novel *The Memorial* we had published in 1932), John Hampson (whose novel *Saturday Night at the Greyhound* we had published in 1931), Rex Warner, and Edward Upward. These two volumes which Roberts edited were, as P. Stansky and W. Abrahams remarked in *Journey to the Frontier*, "taken to mark the beginning, the formal opening of the poetic movement of the 1930s", for they included all the protagonists except one (MacNeice) in that movement: Auden, Spender, Isherwood, and Day Lewis.

When John left the Press in 1932, he went to live in Vienna and he got to know the works and in some cases the persons of the younger generation of writers in Austria, Germany, and France. This gave him the idea of a "magazine in England round which people who held the same ideas about fascism and war could assemble without having to prove their doctrinaire Marxist purity. Why not a magazine to which the writers of *New Signatures* and *New Country* could contribute, side by side with writers like Chamson and Guilloux, and other 'anti-fascist' writers from other countries?"* The project of a magazine had to be abandoned, but in its place Allen Lane and The Bodley Head agreed to publish *New Writing*, a book in stiff covers which was to appear twice a year. Isherwood, Spender, Plomer, Rosamond Lehmann, and Ralph Fox gave their advice and support. *New Writing* was published for some years by The Bodley Head and was then transferred to Lawrence & Wishart. By 1937 John's contract with Lawrence & Wishart was coming to an end and they had "lost interest in *New Writing*",† so that in 1938 he was looking for a new publisher. His entry into The Hogarth Press as a partner

* *The Whispering Gallery* by John Lehmann, p. 232.
† Ibid., p. 309.

solved his difficulties, for we welcomed both him and his ewe lamb, *New Writing*.

We began the publication of *New Writing, New Series, No.* 1 in the autumn of 1938; its title-page announced that it was "edited by John Lehmann with the assistance of Christopher Isherwood and Stephen Spender". Two more numbers were published in the spring and autumn of 1939, but when war came with paper rationing and all its doubts and difficulties for the publisher, it was impossible to continue on the old scale. It did, however, continue in various metamorphoses and under different names until the end of the war. First it became *Folios of New Writing*, shrinking to 159 pages from the 283 pages of the last number of *New Writing*. Later it became *Daylight* in 1941, which became *New Writing and Daylight* in 1942.

These volumes were a cross between a literary magazine and ordinary hard-covered books of short stories, poetry, literary criticism, and politics. They tended, as time went on, to become not only a miscellany, but a miscellaneous miscellany. But at the time when they were first published they were remarkable and valuable. The status of contributors and the standard of their contributions were extraordinarily high. A wide range of British writers of the older and younger generations appeared in the list of contributors, both those who had already appeared in Hogarth Press lists and several new names which have since become distinguished. It is always pleasant to praise famous men and here are some of them who contributed to these volumes: Auden, Isherwood, Spender, Day Lewis, MacNeice, V. S. Pritchett, George Orwell, Henry Green. Before the war cut us off from the continent of Europe many foreign writers leavened the miscellany, some of them well known, e.g. Bertolt Brecht and Jean-Paul Sartre.

The war years were a publishing nightmare for The Hogarth Press, as indeed they were, I suppose, for all publishers. The blackest spot in the nightmare, perpetually preying on our minds, was the shortage and rationing of paper. Having taken John and his ewe lamb *New Writing* into the fold of the Press, we were determined not to allow it to die of starvation and we used a considerable portion of our exiguous paper ration to keep it and its successors going, if diminished in size, until the end of the war. Practically all successful publishers live financially, to some extent, upon books published by them which have become major, minor, or minimal classics. The best-seller, that precarious carrot dangling perpetually before the yearning eyes and nose of even the least asinine of publishers, is extremely exciting and pleasant when one does get hold of one, but few publishers live by best-sellers. It is the books, often very slow selling to begin with, but which establish themselves and go on selling steadily for ten, twenty, thirty and more years, which keep the business rather more than solvent and allow the publisher to sleep peacefully of nights. Virginia's *To the Lighthouse*, which sold in Britain a total of 7,000 copies in its first five years of existence, but in 1967, forty years after it was first published, sold over 30,000 in the year, is a very good example of this kind of book. The Hogarth Press, when the war broke out, had in our list a considerable number of such books which went on selling year after year and had to be continually reprinted. They included all Freud's works, and indeed a large proportion of the psycho-analytical books on our list, all Virginia's books, Vita Sackville-West's, and Rilke's poetry. It was essential, if possible, to keep these books in print, and we had to earmark some of our paper ration for this purpose. We were left with very little paper for new writers and new books. We did manage to do something. We cut down our list drastically; for instance, there are only six books announced in our list for 1941, spring and summer. The standard of what we did publish was, however, pretty high. Between 1939 and 1945 we published Virginia's *Between the Acts*, *A Haunted House*, and *Death of the Moth*. We began the publication of Henry Green's novels with *Party Games* and William Sansom's short stories and novels with *Fireman Flower*. We published books of poetry by Rilke, Robert Graves, Cecil Day Lewis, William Plomer, Hölderlin, Terence Tiller, Vita Sackville-West, Laurie Lee, and R. C. Trevelyan. One of the surprises of the war, which I do not think any publisher foresaw when it broke out, was that, as it dragged on, you could sell anything which could be called a book because it was printed in ink on paper bound in a cover. By 1945, owing to the shortage of paper, and so of books, provided he could print a book the publisher could sell it, apparently in any quantities. What was even more surprising was that one found that one could sell all one's old stock. Novels, biographies, even poetry which had been left high and dry and unsalable before the war were snapped up by booksellers, and apparently by readers, as if they were best-sellers. By the end of the war The Hogarth Press, at any rate, had scraped the last book from the barrel and was left with no unsold stock.

At the end of the war my partnership with John

Lehmann came to an end. I have to deal with this because it had an influence upon the future of the Press and of my life. John has given his account of what happened in the second volume of his autobiography* and I naturally see the facts in a slightly different form, because I saw and see them in a different perspective of prejudice. But apart from the personal question, what happened is, I think, of general and serious importance because our disagreement was fundamentally over the problem and feasibility of the small publisher.

During the war John acted as General Manager as well as a partner. I left him a very free hand, and, considering that we were both what I should call prickly characters, things went on for the most part pretty smoothly. We had two or three rather violent disagreements. They arose because John wished to alter the term of our partnership agreement that no book should be published unless both partners approved its publication. I could not agree to this alteration, because the rule seemed to me essential for two partners running a small publishing business with a list of very carefully chosen books. But that was not really the kernel of our disagreement—the kernel was that John wished to "expand" and I did not. The blessed word "expansion" has been the death of most small publishers. John in his autobiography ingenuously explains the motives for this kind of dangerous expansion. After the war, he says, "I realized that it would be essential to run The Hogarth Press in a different way: to expand in order to carry a proper staff, to have the opportunity to train managers who could take as much as possible of the complex and time-wasting detail off my hands".† But this

* *I Am My Brother*, pp. 310-16.   † Ibid., p. 311.

"expansion" is a euphemism; what it really means is that you have to publish more books in order to pay for an increase in your staff and an increase in your overheads; and because you publish more books, you again increase your overheads and staff, and you then have once more to "expand", i.e. to increase the number of books you publish in order to meet the increase in staff and overheads—and so on *ad infinitum*, or bankruptcy or a "take-over". And behind this circular process is another which John and many another small publisher have ignored at their peril. This kind of expansion entails not only an increasing number of books published and of expenditure on staff and overheads, but also a need for more and more capital. The small publisher who in this way has expanded into a big publisher only too often finds that he may be big, but he no longer controls his publishing business—those who have supplied the capital are now master in his house.

There are in the world of today, I think, two possible ways of publishing books. One is the large-scale business with a large office in central London, a large staff, large overheads, a large invested capital. The iron laws of figures and finance will bind the hands and soul of the publisher: he will have large overheads and he must ever aim therefore at a large and ever larger turnover, which will require a large and ever larger list of books and authors. In this type of business the number of books published by you must be largely determined by the amount of capital invested in your business, the size of your overheads, and the scale of your general expenditure, for it is uneconomic to publish ten books if your business and expenditure are geared to publish a hundred. Naturally the financial urge to "expand", to increase your turnover and therefore the number of books you publish, is powerful and persistent. In a well-established business with efficient directors and an efficient machine the process, within limits, is logical and may be profitable. If you are selling a million cakes of soap at a profit and you increase production to two million, you will probably increase your profit absolutely and proportionally, provided that you can sell the two million. In large-scale business what applies to soap applies to books. But not completely. The trouble with books is that the proviso is much more uncertain than it is in the case of soap. Every cake of soap is exactly the same, but unfortunately for the publisher every book in his list is unlike every other. If you publish 150 books at a profit in 1968, and you add 50 more in 1969 making 200, under the influence of your overheads, you are gambling on the assumption that the additional 50 will sell at least as well and as profitably as the original 150. But, as I have remarked before, the road to bankruptcy is paved with overheads—and books which do not sell. The big, established publishing business, with adequate capital and a long list of successful books still selling regularly, can weather its losses and its overheads. The small "expanding" publisher has no back list of successful "bread and butter books" to balance his losses and overheads and is perpetually harassed by the need to raise more capital. It is not surprising that very few of these small, "expanding" publishing businesses survive expansion.

In his autobiography John writes: "In the late Summer" (of 1945) "I finally decided that Leonard and I had reached a point of no return: if our partnership remained the same, with each of us able to veto any project the other proposed, not only would The Hogarth Press come to a standstill, but my own career would finally be frustrated". I had, in the summer of 1945 and when the war came to an end, no idea that we had come to "a point of no return"—a point and a cliché which temperamentally I am inclined habitually to ignore. We had had, as I said, disagreements, but fewer than, knowing John, I had confidently expected. In the six years of our partnership I had never actually vetoed the publication of a book which John wished to publish, so that the picture of the Press grinding to a halt, with the two frustrated partners unable to agree upon a book to publish, was slightly hyperbolical, if not hysterical.

Although John had decided on this point of no return in the summer of 1945 and had made up his mind to end the partnership, he did nothing for several months and gave me no hint of his intentions. I was therefore extremely surprised to get a letter from him one Saturday morning at the end of January 1946 giving me formal notice that he would terminate the partnership. According to our partnership agreement if either partner gave notice to the other of his decision to terminate the partnership, the partner receiving the notice had an option of buying out the partner who gave the notice. By return of post I formally informed John that I would exercise my right to buy him out.

I received John's letter at breakfast, and, when I had finished my kipper and coffee, I had made up my mind on what I should do about The Hogarth Press. Before lunch I had succeeded in settling its future satisfactorily. I have never been confronted more suddenly and unexpectedly by a major crisis in my affairs and have never succeeded so quickly, completely, and satisfactorily in solving it. Luck was on my side, because, when I saw the possibility of the solution, the means to solve it was almost on my doorstep. To be exact, it was exactly a mile and a half from Rodmell in the village of Iford, in which lived Ian and Trekkie Parsons. In the last three years of the war we had become intimate friends, as I shall relate in a further chapter. In the last year of the war, when Ian was with the Air Force in France, Trekkie stayed with me in Rodmell, and I had helped to negotiate the lease of a house for them in Iford into which they moved as soon as Ian was demobilized.

Ian was a director in Chatto & Windus, the other directors being Harold Raymond, Norah Smallwood, and Piers Raymond. I explained to Ian how John had put his pistol at my head and at the heart of The Hogarth Press, and I asked him whether he and his three co-directors would buy John's share in the Press for the sum which I should have to give John in order to buy him out. My only stipulation was that the Press should retain its independence and not be absorbed in or controlled by Chatto, and that my general policy with regard to the kind of books which we had published and with regard to expansion would be maintained. I would continue to take an active part in what is known as the editorial side of the publishing business; production, sales, distribution, and accounting would be carried out by Chatto & Windus on a commission basis.

Chatto was a moderate-sized publishing firm, but was large-scale compared to The Hogarth Press. It was one of the few remaining big publishing businesses in which the directors seemed to have a policy with regard to books and their publication similar to my own, and there were few, if any, books in our list which Chatto would not have been glad to publish, and vice versa. I felt certain that there would be no disagreement about the kind and quantity of books which the Press would publish. There was no question of a take-over or financial control; the Press was in no need of capital and our common object would be not to "expand", but to maintain the peculiar character, quality, and scale of Hogarth publications. Ian accepted my proposal on the spot and the transaction was put through easily and quickly. Later The Hogarth Press became a limited company, and I became a director of it and, for a time, of Chatto & Windus Ltd.

It is nearly always wrong to believe that events have proved one right, and there is a nasty, smug, ill-conditioned satisfaction in saying: "I told you so", even if it is true that you did tell him so. I did tell John so more than once, and I have the nasty, smug satisfaction of believing that events have proved me right. John went off and started his own publishing business as John Lehmann Ltd. and he proceeded to carry out the programme of publication which, according to him, I had prevented him adopting in The Hogarth Press. His connection with John Lehmann Ltd. lasted for only seven years, and since 1952 he has ceased to be a publisher. This is, I think, a loss both to the art of publishing and to himself. He is immensely energetic; he has a flair for some aspects of the business of buying and selling (not too common among publishers); he has tastes and talents which should have made him a very good publisher. That all these qualities have not brought him the reward which they deserve has been due to two things: he takes life and himself much too seriously, never having learned that nothing, including "I", *sub specie aeternitatis*, matters, and he is much too certain that he is right and the other fellow (even Leonard Woolf) is wrong—a dangerous generalization.

On the other hand The Hogarth Press still exists, having celebrated its half-century of existence last year (1967). I do not think that I over-estimate my achievement or its value, for I do not rate either very high. But in the twenty-three years since John left the Press it has retained its independence and maintained the character, scale, and quality of its publications. One can see this in the books announced in our spring and autumn lists of the years 1955 and 1965, ten and twenty years respectively after John left us:

1955

FICTION

Flamingo Feather by Laurens van der Post

A Contest of Ladies by William Sansom

No Coward Soul by Noel Adeney

The Honeymoon and a Religious Man by Richard Chase

TRAVEL

A Rose for Winter by Laurie Lee

POETRY

Poems 1906 to 1926 by Rainer Maria Rilke

Riding Lights by Norman MacCaig

BIOGRAPHY
Sigmund Freud: Life and Works by Ernest Jones
Letters to Frau Gudi Nolke by Rainer Maria Rilke
Raymond and I by Elizabeth Robins

POLITICS
The Civil Service, edited by Professor William Robson

MISCELLANEOUS
Men and Gardens by Nan Fairbrother
The Dark Eye in Africa by Laurens van der Post
Thomas Hardy's Notebooks by Evelyn Hardy

PSYCHO-ANALYSIS
Clinical Papers and Essays by Karl Abraham
Clinical Papers and Essays on Psycho-Analysis by M. Balint
The Psychology of the Criminal Act and Punishment by
   Gregory Zilboorg
Selected Contributions to Psycho-Analysis by John Rickman

1965
FICTION
A Case Examined by A. L. Barker
Throw by Anthony Bloomfield
It's A Swinging Life by Johannes Allen
Voyage by Laurette Pizer
The Ulcerated Milkman by William Sansom

BIOGRAPHY
Family Sayings by Natalia Ginzburg
Mandate Memories 1918-1948 by Norman and Helen
   Bentwich
Living and Party Living by Jiri Mucha
Apprentice to Power: India 1904-1908 by Malcolm
   Darling

LITERATURE
Virginia Woolf and Her Works by Jean Guiguet
The Collected Essays of Virginia Woolf
Contemporary Writers by Virginia Woolf
Essays on Literature and Society by Edwin Muir
Living with Ballads by Willa Muir

POETRY
Measures by Norman MacCaig
The Year of the Whale by George Mackay Brown

MISCELLANEOUS
The House by Nan Fairbrother

PSYCHO-ANALYSIS
A Psycho-Analytical Dialogue: The Letters of Sigmund Freud and Karl Abraham
Neuroses and Character Types by Helene Deutsch
Collected Papers on Schizophrenia by Harold F. Searles
Normality and Pathology of Childhood by Anna Freud
Psycho-Analytic Avenues to Art by Robert Waelder
The Maturational Process and the Facilitating Environment: Studies in the Theory of Emotional Development by Donald W. Winnicott
Psychotic States by Herbert A. Rosenfeld
The Self and the Object World by Edith Jacobson

For a publishing business which publishes so few books The Hogarth Press has in recent years won a remarkable number of prizes and awards. Miss A. L. Barker is one of the best British short story writers, and she won both the first Somerset Maugham award and the 1962 Cheltenham Festival of Literature Award. We have twice had books which won the W. H. Smith & Son £1,000

Literary Award: *Cider with Rosie* by Laurie Lee, published in 1959, and the third volume of my autobiography, *Beginning Again*, published in 1964.

One of the greatest—and most difficult—achievements of the Press was the *Standard Edition of the Complete Psychological Works of Sigmund Freud* in 24 volumes. We began the publication of this work in 1953 and completed it in 1966. Many years before 1953 I had tried to prepare for an English translation of the monumental German complete edition and had discussed it with Ernst and Anna Freud. The difficulties were so great that for the time being they seemed insuperable and it did not look as if they could ever be overcome. It was obvious from the first that financially the project would be impossible unless we could get the American as well as the British market. But the copyright in the various books was in such a tangled state that it looked as if nothing would ever untangle them. In Britain all Freud's works after 1924 had been published by The Hogarth Press, and I thought that we might be able to come to an arrangement with the publishers of books published before 1924 for their inclusion in the Standard Edition. But the American copyrights were chaotic; some of the books had been poorly translated and we tried unsuccessfully to get the copyright owners to allow us to get them retranslated. In other cases it was doubtful who in fact owned the copyrights. If it had not been for Ernst Freud these dreary problems of law and persons would never have been solved. After the war he went to the U.S.A. and with tact and patience settled the legal and delicate personal questions. This enabled me at last to go ahead with negotiations with the copyright holders to include their property in the Standard

Edition. We had no difficulty with the few English publishers, but in America it was a long and delicate business. At last all was settled successfully and it was possible to go ahead with publication by The Hogarth Press and the Institute of Psycho-Analysis. We took the bold—and eventually profitable—decision not to try to get an American publisher, but to sell the English edition in the U.S.A.

All the 23 volumes—the 24th volume contains the index—were translated by James Strachey. Anna Freud collaborated and he had the assistance of Alix Strachey, his wife, Alan Tyson, and Angela Richards. We—and everyone else—owe an immense debt to James. His translation of these 23 volumes is a marvellous work and rightly brought him the Schlegel-Tieck translation prize in 1966. I doubt whether any translation into the English language of comparable size can compare with his in accuracy and brilliance of translation and in the scholarly thoroughness of its editing. In October 1966 the Institute of Psycho-Analysis gave a great banquet to celebrate the completion of the work, and Anna Freud, James, and I made speeches. I do not find psycho-analysts in private life—much as I have liked many of them—altogether easy to get on with, because they so often cannot conceal the professional fact that they know or seem to know not only what one is thinking, but also what one is not thinking. To stand up in evening-dress and make a speech to several hundred psycho-analysts I found an intimidating experience, partly because they would know (1) what I was thinking, (2) that I was not thinking what I thought I was thinking, (3) what I was really thinking when I was not thinking what I thought I was thinking.

James Strachey died suddenly in April 1967 before the completion and publication of the index in the 24th volume. Today, August 26, 1968, when I am writing this *The Times* reports the death and contains the obituary of his sister Philippa Strachey. She was ninety-six and the last survivor of the ten brothers and sisters whom nearly seventy years ago I first saw gathered in deafening, furious, hilarious argument around the supper table in Lancaster Gate.* I feel I must pause for a moment here to say a word in remembrance of James and Pippa. Practically all the five sons and five daughters of Sir Richard and Lady Strachey were remarkable people. They were nakedly and unashamably intellectuals. Each was a person in his or her own right—a rare thing; they were so individual as to appear strange, eccentric, disconcerting to many people. They were extremely intelligent and amusing; in the realm of ideas they were emotionally violent, but in human relations, though affectionate, they were, I think, fundamentally rather cold and reserved. I knew James when he was a boy at St Paul's and when he came up to Trinity, Cambridge. All his life he was to some extent overshadowed by the greater brilliance, achievements, and fame of Lytton. In similar fraternal cases—not uncommon—more often than not the less successful brother is embittered and, consciously or unconsciously, bears a grudge against his more distinguished brother in particular, and even against the world in general. I never saw the slightest trace of this in James. He was devoted to Lytton and delighted in his success. He confronted the world with a façade of gentle, rather cold, aloofness and reserve, but behind this was a combination of great sense

* See *Sowing*, pp. 190-1.

and sensibility. Unlike Lytton, he had no originality or creativeness, but his editing of Freud shows both the power of his mind and the delicacy of his understanding.

When, as an undergraduate at Cambridge, I first met Pippa Strachey, she was a young woman of twenty-nine. She had her full share of the Strachey mind—extremely intelligent, enthusiastic and highly critical, effervescing with ideas. Unlike her sisters, Dorothy, Pernel, and Marjorie, she was physically attractive and she faced life and human beings with a charming spontaneous warmth which was rare in the Strachey family. I have never known anyone more profoundly and universally a person of good-will than she was, but she was entirely without the congenital vice of so many good-willers—sentimentality. She wore no coloured glasses when she looked at life and people; her attitude was compounded of clear-sightedness, affection, tolerance, amusement, and scepticism. She seemed able to make everything possible and amusing. She once enlisted her friends, including myself, for classes in Lancaster Gate in which she taught us to dance Scottish reels. Though her pupils included such unpromising material as myself and Sir Ralph Hawtrey of the Treasury, it was a great success, and when it was considered that we were proficient, a large dance was given in the Strachey drawing-room at which the crowning point was a display by Pippa's pupils. She devoted her life to women's service as secretary of the society which eventually became the Fawcett Society. In that work she showed that she possessed great administrative ability as well as a strong intellect. With those abilities, if she had been a man, she would almost certainly have attained very high office. That she accepted without complaint the injustice of her own fate, while devoting her life to fighting against the injustices in the fate of others, was an essential part of the charm of her character. It was, too, an essential part in my affection for her.

To return to The Hogarth Press, it has remained a small, independent publishing business with a list deliberately limited to a maximum of round about twenty books a year, though in many years we have published fewer than twenty. We have never published a book for any reason other than a belief that it deserved to be published, a belief which, of course, may have been sometimes mistaken. We have never expanded, never published a book under the financial pressure of expenditure and overheads. We have never been in want of capital, for, as I have explained in *Beginning Again** and *Downhill All the Way*,† the total capital invested in the Press after five years of existence was £136. 2s. 3d., and even that was on account of the printing, not the publishing. By that time the publishing was financed by the profits, and, as the Press has year after year from 1917 to 1968 always made some profit and I have never allowed it to "expand", the problem of finding capital never arose—the Press found its own capital.

John Lehmann in his autobiography implies that he and the war "turned The Hogarth Press into a moderately valuable property".‡ This is an amiable and natural characteristic delusion for which there is no evidence. The Press was probably a more profitable business and a more valuable property before than after the war—this was probably also true of all publishers. I do not think

* Page 253.   † Page 72.
‡ *I Am My Brother*, p. 311.

that the war was good financially for any publisher—
except that it enabled us all to sell unsalable sheets and
volumes; The Hogarth Press, apart from this, suffered
the loss of two of its best-selling authors, Virginia and
Vita. John adds that at the end of the war he thought that
"it would certainly be possible to find the capital for
expansion". But, I repeat, up to that date no one had
really put a penny into the business, other than the
original £136. 2s. 3d. which I spent on the printing-
machine, type, etc.; I never had to "find capital" and
never have since then; the business, because it grew
slowly and successfully, found its own capital. When John
became a partner he did not put any money into the
Press, and when Ian paid me for a half-share in the business the money did not go into the Press as capital, but
into my pocket so that I could buy John out.

In the twenty-three years since John left us the Press
may be said to have pursued the even tenor of its ways not
unsuccessfully. The kind of book it publishes and the kind
of author whom it publishes have not altered. We publish
rather fewer books in 1968 than we did in 1938. This is
mainly due to the fact that, although I still play an active
part and go up to the office about once a week, the active
part is rather passive; i.e. I am content with the good fat
fish in the net or the good young fish who swim into it,
and I no longer go out on the high seas on the lookout for
adventure and the unrecognized genius. This, of course,
is the sclerosis which commonly attacks an established and
successful publisher. The Hogarth Press, by its nature
and history, is peculiarly vulnerable to it. It was from a
business point of view always an anomalous creature,
depending for its birth and existence for many years on

Virginia and me. It remains to a large extent today a personal product, and one result of its happy connection with Chatto & Windus is that inevitably there has come into the business no one who might be the personal successor to me. When I die, therefore, The Hogarth Press will almost certainly also die as an entity. I do not regret this. I deplore the fact that I shall have to die and be annihilated; I should like to live personally for ever. But if I am, as I feel sure I shall be, annihilated, I take no interest in the little odds and ends of me—my books, the Press, my garden, my memory—which might persist for a few years after my death.

I cannot finally leave the subject of the Press and its burial in my grave without returning for a moment to the question whether it would be possible to do today what we did in 1917—create a flourishing publishing business, with no capital and no staff, out of nothing. I am often asked whether I think it possible and I am often told by people with a great deal more experience of publishing than I have that what we did in 1917 was a personal fluke and that today it would be quite impossible for Virginia and me or anyone else to accomplish it. I am not convinced that this is the case. It is true that the business and therefore the art (if there is an art) of publishing have changed enormously in the last fifty-one years; we live, so far as the economic, industrial, and financial system is concerned, in a megalithic age. Everywhere the domination of finance and industry by gigantic companies, with vast capital and enormous turn-overs, becomes more common and more intense. Owing to take-overs and amalgamations publishing is now dominated by big business, it is said, and only these new publishing dinosaurs and megatheriums can hope to make a profit, let alone publish efficiently. Sprats, like The Hogarth Press, have no hope of survival unless they give themselves to be swallowed up by Leviathan, the huge whale or the gigantic shark.

I am not, as I said, entirely convinced by these arguments. It would be more difficult, of course, today to do what we did in 1917. We had some fortuitous and lucky advantages. In Virginia's books we had an enormous potential asset; they were the economic rock upon which in the 'twenties the economic fortunes of The Hogarth Press were profitably based. Many of our closest friends were writers who were to become distinguished or even famous. I do not think it is conceited to say that both Virginia and I were quite good at spotting literary talent or even genius in unknown young writers, and the fact that I was literary editor of the *Nation* during the early years of the Press gave me frequent opportunities to spot them. Again I do not think it is a delusion of conceit which makes me believe that I am a good man of business. I had learned a great deal about the management of an office, about business and finance, in my seven years in Ceylon. As I explained in *Growing**, I owed much to Ferdinando Hamlyn Price, who taught me to be a good man of business. Later in Ceylon my experience for two and a half years as head of a district was invaluable. I was responsible for the revenue and expenditure of the district and for the accounts. And I knew what I was doing as an accountant, for I had had to pass an examination in accounts before I could be promoted to administer a district. As Assistant Government Agent of Hambantota, in

* Pages 106-11.

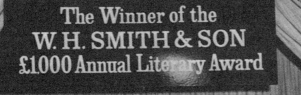

The author making his acceptance speech for the W. H. Smith Award
John Betjeman and David Smith are at left.

addition to the ordinary business connected with Government revenue and expenditure, I was responsible for running a fair-sized industry, the manufacture, sale, and distribution of salt, which was a Government monopoly. I am, by nature, a good businessman, and up to a point enjoy administration and organization and accounts and figures and dealing with all sorts and kinds of people. After the administration, business, and finances of a Ceylon district, running The Hogarth Press seemed to be child's play and a spare-time occupation. The knowledge of business which I gained from my experience in Ceylon was thus of great value to us in the early days of the Press, particularly when events forced us to allow it to develop into a serious publishing business. It enabled me to understand and control its finances and so consciously to adopt the policy of limiting its operations and so resisting the fatal lure of expansion.

The advantages enjoyed by us, which I have just described, were, of course, the foundation of the development and success of The Hogarth Press. It would be rare for anyone at any time who wanted to start as a publisher to find that he had up his sleeve such, and so many, good writers as Virginia and the other authors whom we published during the first five years of The Hogarth Press. Let me recall their names in the order in which they appeared on our lists from 1917 to 1922: Virginia and Leonard Woolf, Katherine Mansfield, T. S. Eliot, J. Middleton Murry, E. M. Forster, Hope Mirrlees, Logan Pearsall Smith, Gorky, Bunin, Dostoevsky, and in the next two years we published books by Roger Fry and V. Sackville-West and the *Collected Papers* of Freud. This is a remarkable constellation of stars, a formidable list of publications. It is also probably rare for intellectuals like myself to take the trouble to become good businessmen. But I am not so foolish as to believe that our advantages could not occur again. There is no reason to believe that it is impossible that tomorrow or tomorrow or tomorrow there may not be a circle of young, unknown, brilliant writers whom someone might begin to publish on a small scale as we did in 1917. And there is no reason why he should not succeed as well as we did, provided he is a good businessman and is determined to limit his operations, refusing to listen to the John Lehmanns singing their siren song about expansion which can lure one so easily on to the Scylla of the take-over or the Charybdis of bankruptcy.

Chapter Three

1941-1945

AFTER Virginia's death I continued to live in Rodmell. Many of my friends, I think, felt that I ought not to remain alone there; they offered to come and stay with me or asked me to stay with them. It is no good trying to delude oneself that one can escape the consequences of a great catastrophe. Virginia's suicide and the horrible days which followed between her disappearance and the inquest had the effect of a blow both upon the head and the heart. For weeks thought and emotion were numbed. My mind was haunted by certain phrases of Claudio in *Measure for Measure:*

> To lie in cold obstruction and to rot

and

> to reside
> In thrilling region of thick-ribbed ice.

They applied, not to the dead, but to the living, to me. I remained where I was, for, in fact, there was nothing else to do. You cannot escape Fate, and Fate, I have always felt, is not in the future, but in the past. I have my full share of the inveterate, the immemorial fatalism of the Jew, which he has learned from his own history beginning 3,428 years ago—so they say—under the taskmasters of Pharaoh in Egypt, continuing 2,554 years ago in the Babylonian captivity when Nebuchadnezzar was king in Babylon, and so on through the diaspora and the lessons

127

of centuries of pogroms and ghettoes down to the lessons of the gas chambers and Hitler. Thus it is that we have learned that we cannot escape Fate, because we cannot escape the past, the result of which is an internal passive resistance, a silent, unyielding self-control.

So I continued to live in Rodmell. I had nowhere to stay in London, for bombs had made the house in Mecklenburgh Square uninhabitable. The scaffolding or skeleton of my life remained the same. Work is the most efficient anodyne—after death, sleep, or chloroform—for pain, whether the pain be in your great toe, your tooth, your head, or your heart. To work and work hard was part of the religion of Jews of my father's and grandfather's generations. In the ghetto for hundreds of years, I imagine, you had to work hard to keep alive, and long before that when Adam heard the voice of the Jewish God walking in the garden in the cool of the day, he heard it say: "In the sweat of thy face shalt thou eat bread, till thou return unto the ground; for out of it wast thou taken: for dust thou art, and unto dust shalt thou return". I doubt whether there is any great difference in the genes and chromosomes of the various tribes, races, and nationalities that have inherited and desolated the earth, but their ways of life, their laws and traditions and customs, the fortuitous impact and the logic of events and history have gradually moulded the minds and characters of each so that often they differ profoundly from one another. There is, I think, or there was, a tradition consciously or even unconsciously inculcated in Jews that one should work and work hard, and that work, in the sweat of one's brain as well as of one's face, is a proper, even a noble, occupation for all the sons of Adam. I think that my father had absorbed this tradition and instinctively obeyed it, and that, young as I was when he died, I had observed it and again, in my turn, instinctively obeyed it. In the ordinary sense of the words, I doubt whether I have a "sense of duty", but I have always felt the urge, the necessity, to work and work hard every day of my life.

I worked immensely hard at school—both at my prep school and at St Paul's—and there for the first time I came up against a tradition exactly opposite to mine with regard to the ethics of work. It was at school that I first, and very soon, heard the word "swot"—"he's a bloody" or "he's a dirty swot". The word "swot" derives from the word "sweat", and so appropriately carries us back to the voice of the Jewish God walking in the garden in the cool of the evening. The tradition of the English public school was in my youth, and had been for a hundred years and more, the aristocratic tradition which despised work and the worker. It had spread all through the educational system. Even the masters at my prep school in Brighton and at St Paul's—most of them had been themselves educated in public schools—despised the dirty swot and often showed their contempt for him. It was only cricket and football, not work, which a gentleman took seriously. I very soon observed the difference in values between my attitude to the swot and that of the other boys and the masters in my prep school in Brighton when I went there in 1892 at the age of twelve. I kept my knowledge to myself, but, being by nature stiffnecked and pigheaded, I went my own way and worked tremendously hard all through my school days. Being fairly good at games, I escaped much of the odium of being a swot, until, when one reached the top forms in St Paul's, one was high enough in the establishment to be able to work without indignity.

During the remainder of my life, after school and the university, I have never ceased to work long hours and intensively. In Ceylon I normally worked a twelve-hour day, and in the six and a half years that I was there I had —apart from the weeks when I was ill with typhoid— only about four weeks' holiday. In the fifty-seven years since my return to England I have worked no less hard and persistently. I do not claim this as a merit, but state it as a fact. Both Virginia and I looked upon work not so much as a duty as a natural function or even law of nature. Except when we were officially "on holiday", we each every day retired to our respective rooms and worked from 9.30 to 1, and it was as natural and as inevitable that we should do so as that we should go to bed and sleep each night. In London in the afternoon almost always I worked in the Press or on some political committee, and if we were in in the evening, I would usually read a book for review or in connection with what I was writing. Virginia's working day was as long as, perhaps longer than, mine. What she was writing or going to write was rarely not in the centre of her mind. She was continually thinking about the book she was writing or was going to write, or she was living, observing, or absorbing the raw material for them. She also did an immense amount of reading either for her essays or for reviewing; she read the books intently and intensively and there are still in existence an immense number of notebooks full of the notes which she made methodically as she read. I should say that in an ordinary normal day of twenty-four hours we each of us slept for eight and worked ten or twelve hours.

I continued to work even harder and longer after Virginia's death. I was writing *Principia Politica* with unconscionable slowness. Then there was The Hogarth Press; I had an almost daily correspondence with John Lehmann and frequently met him in London for discussion and decisions, and I occasionally went to Letchworth to see the staff there. In 1931 I had become joint editor of the *Political Quarterly* with Willie Robson; when the war came, Willie went into the Ministry of Fuel and Power and had to give up his editorship. From 1940 to 1946 I was sole editor and, under war conditions, this was no easy business. A more or less "learned" or rather expert journal like the *Political Quarterly* presents a completely different editorial problem from a weekly like the *Nation* or *New Statesman* or, I should think, still more a daily paper. The trouble which faces the editor of the weekly every Monday morning is that he has too much material and too little space; the nightmare which perpetually haunts the editor of the quarterly is that he will find himself with too much space and too little material. This is mainly due to technical difficulties. Three months is an interval of time which is more likely to make an article out of date than seven days, and an editor of a quarterly who postpones publication of an article which was written for a January issue until the April issue will be publishing it nearly six months after it was written. He has therefore to plan each issue more meticulously and spatially exact than the editor of the weekly. Looking through these numbers of the *Political Quarterly* over twenty years after I planned and produced them, I feel a certain parental pride in them, and I do not think it is merely parental infatuation which makes me think that journalistically the standard is remarkably high. I find them readable even today, though one must allow for the fact that almost every paper or journal is more interesting twenty years than it was twenty minutes after it was published. Most of the articles are written by intelligent experts—not a very common combination—and deal with events and problems, even contemporary events and problems, of political or sociological importance from the long-term, if not eternal, angles of incidence and reflection. The mere Contents of two issues, and the names of the writers, which I give below, are some evidence of this:

October-December, 1942

German Disarmament and European Reconstruction	by Mercator
The Meaning of the French Resistance	by Professor Paul Vaucher
Hitler's Psychology	by Leonard Woolf
Colonies in a Changing World	by Julian S. Huxley
Industry and the State	by Joan Robinson
"Grey Eminence" and Political Morality	by the Hon. Frank Pakenham
Christianity, Science, and the Religion of Humanity	by Anceps
Putting Britain Across	by Historicus
Parliament in Wartime	by H. R. G. Greaves

April-June, 1943

The Problem of the Public Schools	by R. H. Tawney
The Beveridge Report: an Evaluation	by William A. Robson
The Age of Transition	by Harold J. Laski
Parliament in Wartime	by H. R. G. Greaves
The Future of An Alliance	by Max Beloff

Another of my occupations was, every now and again, to sit on the Civil Service Arbitration Tribunal. As I recorded in *Downhill All the Way*,* in 1938 I was appointed a Member of the National Whitley Council for Administrative and Legal Departments of the Civil Service. What these many words meant was that Civil Servants were precluded from striking, and, if their Trade Unions and H.M. Treasury could not agree with regard to any claim about pay or other conditions of employment, the claim had to be remitted for trial and decision to a Civil Service Arbitration Tribunal. The Tribunal consisted of three arbitrators: a permanent chairman, appointed by the Government, an arbitrator from a panel appointed by the Treasury, and an arbitrator from a panel nominated by the staff or Trade Unions. I had been nominated by the staff and continued so to be for seventeen years. I used to enjoy my work as a District Judge or Police Magistrate when I was a Civil Servant in Ceylon, and I found my work as a Civil Service arbitrator in London hardly less interesting. To sit on the Bench and try almost any case as judge, magistrate, or arbitrator (provided that you think about the litigants and their case and characters, and not about yourself—a rule not always observed by judges) can give one an insight into the mind, motives, and methods of the human animal which it is not easy to obtain in any other way. To be above the battle, and to know that you must remain there and under no circumstances allow anything—even the shape of a litigant's nose or the colour of her eyes—to "prejudice" one, purges the mind and purifies the vision so that one sees things and understands people in a way impossible in everyday life.

Even when the skies are falling about our ears—

* Pages 207-17.

whatever may be happening to Justice—we eat our eggs and bacon for breakfast and go about our daily and nightly business. With German bombs falling about our ears in Regent's Park or with battles raging from Normandy to Rome to decide the fate of the world, the Arbitration Tribunal met from time to time throughout the war to decide such questions as whether a clerk in the Ministry of Labour should get an increment of 1s. or 2s. or whether a Prison Matron should work forty-four or forty-five hours a week. This is, I think, as things should be and indeed always have been: the ordinary person, except when he was being killed, starved, conscripted, or ruined by the great World Wars and the great World Conquerors and Pests of the World like Alexander the Great and Napoleon, has ignored them and got on with his eggs and bacon, or, like Jane Austen, ignored Napoleon's retreat from Moscow and got on with writing *Mansfield Park*.* So far as I was concerned, when the great battles which were to decide the great war were at their height in Poland, Hungary, Greece, Italy, and France, and the "doodle-bugs" or V.1's were falling on London, for in-

* See Jane's letter to Cassandra written in Chawton, Sunday evening, January 24, 1813, in which she says that she learns from Sir J. Carr that "there is no Government House in Gibraltar", as she had said in *Mansfield Park*, and that she must alter it to the Commissioner's. There is no mention of the Great War, or of Napoleon, or of the retreat from Moscow, but she does write: "My mother sends her love to Mary, with thanks for her kind intentions and enquiries as to the Pork and will prefer receiving her share from the two *last* Pigs: she has great pleasure in sending her a pair of garters, and is very glad that she had them ready knit". How right that the great writer should confer immortality, not on the great conqueror and his great war, but on Mary, the Pork, the two *last* Pigs, and the pair of garters.

stance, I spent a day in December, 1944, in the house looking on to Regent's Park, with Sir David Ross and Mr Fairholme, deciding whether Chief Officers, Matrons, Superintendent of Weaving and the Superintendent of Printing and Binding in the Prison Service should be entitled to payment at time-rate-and-a-quarter for all hours worked in excess of eighty-eight a fortnight. And after listening for many hours to a recital of the hours and conditions of employment and of the scale of salaries, and to the arguments for and against by the Prison Officers' Association and the Prison Commissioners, we awarded the claimants what they asked for for a period of three years.

There is one thing I should like to say before finally leaving the Civil Service Arbitration Tribunal. For me personally and psychologically, as I have said, I always found the work interesting, though at the same time I always felt that from a public point of view the whole system was absurd and a waste of time. As I pointed out in *Downhill All the Way*,* the whole industrial structure of the Civil Service seemed to me crazily irrational. Each of the hundreds of Government occupations has, on the face of it, a scale of pay and conditions of employment peculiar to itself, but in fact, of course, in any particular case there will be close similarity with a large number of other occupations. Hence, if any change is made in the scale of pay or conditions of work in one occupation, it sets off an unending chain of claims in similar occupations throughout Government service. I said that I thought that the whole structure of Government service should be rationalized by constituting a small, limited number of classes for scales of pay and conditions of employment, and this

* Page 214.

THE JOURNEY NOT THE ARRIVAL MATTERS

classification and pay structure should be applied to practically all persons in Government employ. I am glad to see that the Fulton Commission on the Civil Service, which has recently reported, makes the same proposal.

Sporadically, from time to time and for short periods of time—a few days or a week or two—I used to edit the *New Statesman*. This was because the editor, Kingsley Martin, when he wanted to go on holiday for a week or more or when he went on one of those peregrinations in Europe or Asia which are essential stimulants to the life-blood of good journalists, had got into the habit of asking me to act for him. Thus in 1943 I did eight weeks in all, though I stipulated that I would come for only two or three days in the week. It was a pretty strenuous business fitting it in with my other work, but I must admit that, knowing that I would never be a permanent prisoner in the editor's chair, I got a good deal of amusement out of it. The staff which I had to deal with consisted of Dick Crossman, G. D. H. Cole, Aylmer Vallance, Norman Mackenzie, and on the literary side Raymond Mortimer.* I found them easy to get on with, but they were a formidable team and required some watching. We met together on Monday or Tuesday for the final decision on what the menu should be for the paper to offer for the week. I always contemplated Dick Crossman with amazement and the greatest admiration. He was the best journalist I have ever known. His mind was extraordinarily fertile of ideas; it teemed with them, and if you dipped into it, you

---

* It was only after the war that I had Dick as Assistant Editor. He was in fact Assistant Editor of the *New Statesman* from 1938 to 1955, but during the war he was in the army a brilliant Director of Psychological Warfare.

brought up a shoal of brilliant, glittering ideas, like the shoal of shining fish that one sometimes sees in a net pulled out of the sea by a fisherman. It is true that Dick's ideas were almost as kaleidoscopic in colour and as slippery to keep a hold on as the mackerel for, having written a glittering and devastating article one week, he would turn up the following Monday with the most brilliant idea for the most brilliant article contradicting his most brilliant article of the previous Monday. And on each of the two Mondays Dick, I am sure, believed passionately in each of the two ideas.

The rest of the staff were journalistically an almost perfect team. Douglas Cole was a very old friend; for years I had worked closely with him and Margaret, his wife, in the Fabian Society and the Labour Party. He was an extraordinarily able man. Academically, as a teacher in Oxford, and politically, as an intellectual providing the British Labour Movement and Party with ideas, principles, and policy, he had a large and devoted following, particularly among the young. To an editor he was what every editor prays for, being as reliable as the sun and moon. For one could be absolutely certain of receiving by the first post on Wednesday morning an impeccable article, of exactly so many thousand words, on one of those topical, but grimly gritty subjects, which are the despair of editors—and often of readers—for they lie in the depressing region where economics, industry, trade unionism produce the most important, insoluble, and boring problems.

Norman Mackenzie, when I first met him in the *New Statesman* office, was a young man at the beginning of what one thought (wrongly) would be inevitably a journalistic career. He was as reliable as Douglas, never putting a foot—or a pen or typewriter—wrong. He covered much the same field as Douglas; it is a field in which articles have to be written by "experts" and, by their nature, do not make light or amusing reading. Sitting in Kingsley's chair, there was only one editorial grumble which I had when I read the articles of Douglas or Norman. They were uncompromisingly sound and solid, but verbally they were written by plain cooks. It was only when, after the war, Dick reappeared that one realized that it was possible for an article, even about the League of Nations or Ernest Bevin, to be sound and solid and yet at the same time brilliantly readable.

The last member of the *New Statesman* team was Aylmer Vallance. He was very much a professional journalist, for he had been the editor of the *News Chronicle*. During the war he combined his work on the *New Statesman* with that of a Colonel on the General Staff at the War Office. In the technique of planning, writing, and preparing for press a weekly paper he was first class; he could knock off in an hour a good (but not a very good) article on any subject from mind to matter or from God to girls. He was a good fellow well met. But he was also one of the most indiscreet good fellows that I have ever met. For instance, one day in a room full of people going and coming, as they do in a newspaper office, he told us that he had been interrogating captured German officers and that they were all convinced that Germany would win the war with a new weapon Hitler was about to use against us —and he described in some detail what a little later we came to know as the V2. Hearing this from an officer in a colonel's uniform with red tabs and all, in a room full of heterogeneous people, it never struck me that there could be anything wrong about it. But later in the day when I met Bunny Garnett, who was in the Air Ministry, and asked him about Hitler's new weapon, hair and hackles rose upon his head and he told me furiously that I had no right to be in possession of—far less talk of—what was a top top secret—only that morning—with metaphorically all doors shut and blinds pulled down—for the first time the facts about this topmost secret had been revealed to the topmost red tabs of his office. I felt that I had only just escaped arrest and imprisonment.

This was not the only time I got into trouble through Aylmer. He was distinctly a Fellow Traveller, and may, for all I knew, have been a member of the Communist Party. I had lived long enough in the Fabian Society and the Labour Party, among denizens of the political Left, to know that you could never completely trust a Fellow Traveller, that dear friend who might or might not be a crypto-communist. But it was some time before I realised that one had to keep more than one eye on Aylmer. It was in May 1945 that my two eyes opened. I edited the paper during the four historic weeks in which Hitler's death and the end of the war with Germany were announced. It was a hectic time. After my third week of editing, I received an enraged letter from Maynard Keynes about the front page article celebrating the end of the war. I cannot now remember what the article was all about, but I think it must have been full of the slants, snides, sneers, and smears which Communists and Fellow Travellers habitually employ as means for building a perfect society. Maynard was outraged, and so was Lady Violet Bonham-Carter and several other highly respectable and respected persons who had written indignant letters to him. For some strange reason Maynard blamed the absent and innocent Kingsley instead of me. I wrote him the following letter in order to divert his anger on to my guilty, but not too contrite, head:

Dear Maynard,

    I entirely agree about the article, although the responsibility is of course mine. It is appalling. I was in a difficult situation last week and I daresay looking back I made a wrong decision. The two days' holiday in the middle of the week meant that all the proofs had to be passed on Thursday. I arranged that Vallance should write the front page and leave it with the printer on Thursday morning so that I could not see it before it was in proof. . . . It was understood that the printers could only print on Friday if there were no serious alterations to be made. The front page was only ready for me to read at 5. When I read it, I felt as you do about it. . . . It was considerably worse than it is now. The difficulty was that the article ought to have been rewritten entirely, but that in the state the printers were it meant beginning all over again on Friday for them and quite probably not getting the paper out until Monday. I also had an engagement which made it necessary for me to catch the 6.45 latest to Lewes. In the end I told Vallance that he must put in certain alterations and additions which I thought would make the article tolerable. But I agree that they did not and that it would probably have been better to have re-written the article and have held up the printing.

    As regards what action it is your duty to take, I think

it would be wrong to confuse this case with anything against Kingsley. It should be raised at the Board meeting, but the responsibility is mine and it shouldn't be counted against Kingsley.

I don't understand Vallance. Up to this incident I had always thought him to be a first rate journalist and second rate in everything else, but also someone one could trust to be reasonable up to a point.

Yours
Leonard

I think this incident and my letter gives a good idea of the hurry and scurry of editing a paper like the *New Statesman*. The scene described by me to Maynard took place, not in the *New Statesman*, but at the printers in Southwark. There one had to go on Thursday morning—in this case on Friday owing to the holidays—and pass the proofs, or sometimes write or rewrite the article on the front page. One sat in a kind of glass case and the page proofs came up to one straight from the machine. Usually it was merely a question of correcting the proofs, but sometimes something might have happened overnight to make the article written the previous afternoon now out of date. In that case one might have to do a great deal of rewriting or even write an entirely new article. One worked under high pressure with all sorts of comings and goings, with the printer metaphorically behind one's back clamouring for the copy. It was much the same turmoil, or even more so, all the week from Monday to Wednesday in the office in Holborn. I do not mind working under pressure and am not disturbed by disturbance, having to do two or three things at the same time with doors

141

perpetually opening and shutting and people perpetually coming and going. Ceylon had taught me to work in its kachcheries, unprotected by doors or windows, impassively and imperturbably in a kaleidoscope of noise and perpetual motion, people talking in two or even three languages at the same time about two or three distinct questions at the same time. There is, in fact, a certain exhilaration in this kind of expertise, the administrator, journalist, or tycoon able to juggle with half a dozen problems at the same time like a juggler who can keep half a dozen billiard balls in the air at the same time. But to do this kind of thing all the time for a long time has a curious effect upon the mind. You live on the surface of things, on the surface of life and its problems, on the surface of your own mind. You become so slick, so skilful and astute, so knowing, that you no longer need to, or eventually can, think; you know all the questions and fortunately—or unfortunately—all the answers. I learned in Ceylon that if you have to settle any question which really requires some thought, whether in a Government office, a publisher's office—or even a newspaper office— it is essential to take it home with you. In your office you have bright ideas which seem to you brilliant. It is only at home that, if you ever think, you may think.

I have already in *Downhill All the Way** described what seems to me to be the effect of journalism, if persisted in for long, upon the mind of the journalist, and I will not repeat it here. There is, however, one rather interesting psychological effect of journalism upon journalists which I did not mention there, but which I noticed when I was Literary Editor of the *Nation*, and

* Pages 140-1.

again when I temporarily edited the *New Statesman* for Kingsley. All occupations or professions, like individuals, create around themselves a kind of magnetic field. To me myself everything within and without myself acquires a curious and strong quality or aura of me myself—my pains and pleasures, my typewriter and my big toe, my memories and the view which I am now looking at from my window, the people I love and the people I hate, all these, when they enter the magnetic field which my ego and egocentricity have developed about me, acquire a meaning and value peculiar to myself. And everyone else walks through life, materially and spiritually, enveloped in a similar magnetic field of his own personality which gives to everything and everyone entering the field a magnetised reflection of his ego, a meaning and value which he alone in the world feels and understands.

Occupations and professions, even institutions, acquire the same kind of magnetization. Everything entering the magnetic field surrounding a school or college, the occupation of a barrister or doctor, of a miner or electrical engineer, a cook or a gardener, acquires the same kind of peculiar meaning and value to those within the field. The psychology of this occupational hallucination or self-deception is shown most obviously and commonly in the enormous, sacred importance which the vocation and everything connected with it acquire in the eyes of those who practise it. Kings and queens, their families and relations, and all those who earn their living by some sort of Court service, have always reached the most fantastic heights of ludicrous hallucination, and they have been encouraged by the almost universal acquiescence of ordinary people whose passion for self-deception is so great and so deep that they are delighted to be deceived even by the self-deception of someone else. Judges and priests —particularly the higher classes, the Popes, Cardinals, Archbishops, and Bishops—come second to kings and queens in the quality and quantity of their vocational inflation and self-deception. I think journalists come third. I am thinking, of course, primarily of the daily paper and the intellectual weekly. The chief factor in the hallucinatory overestimation of the importance of journals and journalism by journalists is the obvious importance of the events and subjects on which they are daily or weekly pronouncing judgment anonymously *ex cathedra*, the cathedra being in fact the editor's chair. By the curious logic of history and human institutions, the Pope, a celibate virgin who is forbidden to have any relations with women, is entrusted with the power to make detailed and intimate regulations for millions of ordinary people regarding marriage and the sexual intercourse of husband and wife. Not unnaturally a man who is given the power to make infallible decisions with regard to such important matters claims successfully from millions of people an enormously inflated importance, the outward and visible sign of which is the fantastic fancy dress in which, as with queens and kings, he is habitually photographed.* Something of the same sort happens to the editor, the paper

* The credulity of human beings is so gigantic and unquenchable that millions of them not only accept the dictates of an old gentleman in Rome about contraceptives, but also believe that he is in direct communication with the Deity who created the universe, with its suns and galaxies and comets flaming through infinite space, and that it is direct from this Deity that he, the Pope, has received the detailed instructions with regard to how married persons are permitted to use or not to use contraceptives.

which he edits, and all those engaged in producing the
paper. I think that they all—even the lowliest office
girl—feel that the paper is important because it is daily or
weekly pronouncing judgment on the most momentous
events, persons, and policies which history causes to pass
like a pageant from Monday to Thursday morning, when
the paper goes to press, before the editor's and the assistant editor's desks. The competence of the editor to
pontificate on some of these subjects is probably no higher
than that of a celibate old gentleman in Rome to lay down
the law on the intimacies and intricacies of copulation and
the mechanics of contraceptives. But it is quite impossible
not to believe that one is important if one is perpetually
laying down the law upon important questions. And with
journalists, as with Popes, judges, and M.P.s, the power
complex also comes in. Every editor—certainly every
good editor—believes, not only that he is continually pronouncing judgment about the most important questions
but that he and his paper have a powerful influence upon
public opinion with regard to those questions. Thus a
magnetic field of highly charged importance, influence,
and power is created around every newspaper, and everyone connected with it is subjected to its effect and to any
vocational delusions to which it gives rise. I know from
experience that the moment I sat down in the editor's chair
in the *New Statesman* office, though I am by nature
sceptical, an unusual sense of importance, a tinge of *folie
de grandeur*, enveloped me. It emanated from the magnetic
field of the *New Statesman* into which I had suddenly and
importantly entered. Instinctively I was feeling that
everything I was going to do or say during the next week
was of importance. I was the (temporary) wielder of

influence and power. I used to feel the same thing in the *Nation* office when I was Literary Editor there, and even as editor of the *Political Quarterly*. And this effect of the newspaper's magnetic field extends, as I said, far beyond the editor's chair. I am sure that Maynard Keynes, for instance, would never have taken so serious a view of Aylmer Vallance's indiscretions if he had not attributed such immense importance and influence to what appeared in the *New Statesman*.

In the last sentence I very nearly wrote "exaggerated" instead of the word "immense". The question of whether or to what extent one exaggerates one's own importance and that of one's work or productions is a painful one and not entirely easy to answer accurately. I feel pretty certain that the magnetic field surrounding journalism induces the editor and staff of every newspaper to believe that his paper is much more important and influential upon public opinion than it really is. I have no doubt at all that I was a victim of this occupational delusion as editor of the *International Review*, the *Nation*, the *New Statesman*, and the *Political Quarterly*. Most of this is, of course, conjecture, but the evidence of facts, so far as it goes, seems to show that newspapers have very little positive effect on the formation of opinion. This is certainly true of the millions of copies of the popular dailies churned out and sold under the hourly direction of the great Press Lords, Northcliffe and Beaverbrook, both of whom aimed at moulding public opinion on almost any subject from sweet peas to Empire Free Trade, and seemed to believe—quite wrongly—that they succeeded. Some of their modern successors suffer from the same illusions. The truth is that an enormous majority of newspaper readers read them for either one or both of two purposes. The first purpose is simply to learn what has happened, the facts, whether about racing and football, crime and sex, the doings of the Queen and her family, or politics. The second purpose is to obtain entertainment, pleasure, reassurance, or irritation. Clearly an enormous number of people read papers mainly to get entertainment or amusement from them. A smaller number want to find in them a confirmation of their own tastes, beliefs, loves, hatreds, and delusions—they want to be reassured. A still smaller number read them because they want to be annoyed. This applies particularly to the intellectual weeklies. I am sure that many people have always read the *New Statesman* because it supplied them with a weekly dose of justified irritation. Finally I repeat what I have said elsewhere that I think it highly probable that the influence of newspapers is in inverse proportion to the magnitude of their circulations. The millions of copies of the *Daily Mail* or *Mirror* and the millions who read them are so formless and fluid that the papers have practically no effect upon the minds of the readers. Journals with a very small circulation, written by experts for experts on more or less technical subjects—the *Political Quarterly* is one of them—more often than not are dealing directly with opinion and are more likely therefore to influence it.

These facts about the magnetic field surrounding occupations and the occupational delusions which are involved in it lead to a question which almost everyone, I think, as he grows old, particularly if he writes his autobiography, must occasionally ask himself—and the autobiographer in particular must try to answer it honestly. Here I sit at the age of sixty, seventy, eighty, or (in my own case today) eighty-eight; behind me lies "work", anything from forty to seventy years of "work". I am talking of males of the middle class; we began to "work", to go into a profession or a business some time between the ages of eighteen and twenty-four. We had gone to prep schools (for seven years) and public schools (for five years) working to prepare ourselves for the work which we were going to do in our profession or business. What was the object of this "work", of these hours and years of labour? What did we think was its object? What did it achieve and what did we think that it achieved? Of course, owing to the economic determination of history, classes, and individuals, we worked in order to make a living, and, since we are alive to ask these questions, we presumably achieved it. But, while there is a profound truth in Marx's analysis of society, psychology, and economics, it is only a half or possibly even a third of the truth. The vast majority of human beings regard their work, not only as economically determined, i.e. a source of income, but also as having a non-economic object and value producing effects of social, psychological, or artistic value.

It is sixty-four years since the November day when I set sail from Tilbury through fog and drizzle down the Thames in the P & O ship *Syria* for Ceylon. It was the beginning of my work in the technical sense, work in a profession, work to earn a living. In the sixty-four years which have passed since that November day I reckon that at a minimum I have worked 158,720 hours or the equivalent of 6,613 days. I feel sure that this is an underestimate, that I have, in fact, during the last sixty-four years spent on what everyone would agree was "work" considerably more than 170,000 hours. The kind of work

which I did in the seven years which I spent as a civil
servant in Ceylon I have described in *Growing*; the kind
of work which I have done since I resigned from the
Ceylon Civil Service I have described in *Beginning Again*
and *Downhill All the Way*. In order to get some idea of
its object and effect I propose to examine what exactly
this work consisted of during the six years of the war,
1939 to 1945, at the end of which I was sixty-five years old.

The routine of my life changed considerably from time
to time during those six years; the changes affected the
amount of time which I could give to my various occupations, but not, I think, the total amount of time which I
gave to work. What chiefly determined the kind of work
which I could and did do was the proportion of my time
which I spent in London and in Rodmell. Virginia's
death which disrupted the whole of my life disrupted the
rhythm and routine of my work; but what I could do and
how I could do it was enormously influenced by the
bombing of London. I have already described how in
1940, before Virginia's death, the early bombings
wrecked our house in Mecklenburgh Square and made it
uninhabitable. We now had nowhere to stay in London
and became for the first time in our lives country folk,
living permanently in Monks House, Rodmell. The first
effect of this and of the evacuation of The Hogarth Press
in September 1940 to Letchworth was that I could no
longer take any part in the day-to-day control of the Press;
my work as a publisher was reduced to remote control by
correspondence with John Lehmann and by our meeting
from time to time in London or Rodmell. I went up for
the day to London whenever I had work to do in the
Labour Party, Fabian Society, Arbitration Tribunal, or

New Statesman; otherwise I sat in Rodmell writing or editing the *Political Quarterly*.

But after Virginia's death I felt that I must have somewhere where I could stay in London in order to be able to do my work there more intensively. So I took a flat in Cliffords Inn. I discovered that by nature I am not a flat-dweller; I have little or no sense of gregariousness; I find no comfort or security in the sound and smell and warmth of the herd, the coziness of the human rabbit-warren. I like my fellow-human beings, but I require considerable periods of absence from them, periods of silence and loneliness. I could not stand Cliffords Inn for long, and in April 1942 I got three rooms in my house in Mecklenburgh Square patched up and moved in there. "Patched up" is the right description of the rooms and the house. There were no windows and no ceilings, and nothing in the house, from roof to the water pipes, was quite sound. I got my loneliness and my silence (except when the bombs were falling) all right. But I have experienced few things more depressing in my life than to live in a badly bombed flat, with the windows boarded up, during the great war. I stuck it out in Mecklenburgh Square for exactly a year, but by October 1943 I could bear it no longer and I took a lease of 24 Victoria Square. When that lease came to an end I bought a further ninety-nine-year lease of the house from the Grosvenor Estate, so that, if I live to the age of one hundred and fifty, I may still have a house in London.

Cliffords Inn, Mecklenburgh Square, and Victoria Square, in succession, made a new routine and rhythm of life for me from the winter of 1941 to the end of the war in 1945. I began again to do a good deal of political work in London. The routine which gradually established itself was two to four nights in London and the rest of the week in Rodmell. Looking through the list of my engagements I find that in the last years of the war my "work" in London, most of it political, consisted of the following:

Labour Party: Secretary of the Advisory Committee on International Relations; Secretary of the Advisory Committee on Imperial Questions.
Fabian Society: Executive Committee; Chairman, International Bureau; Imperial Bureau.
Anglo-Soviet Society.
New Statesman: Board of Directors.
Civil Service Arbitration Tribunal.
Political Quarterly: Board and Editor.

This "work" took up many hours of my time, for, when in London, I often had two committees in a day and, quite apart from the committees, I often wrote reports for the Labour Party and for the Fabian Bureaux. All this work was unpaid, except the *Political Quarterly* and the Arbitration Tribunal. (My salary for editing the *Political Quarterly* was £80 and I was paid four guineas for each case which I heard on the Tribunal.)

Why did I do all this work year after year? I was sixty-five in 1945 when the war ended and sixty-five is the usual year for retirement. But I continued to grind away at much of this political or social work for many years after the end of the war. And even in Rodmell I did a lot of work of the same kind, of a political, social, or communal nature; for I was Clerk to the Parish Council for seventeen years, was for years and still am a Manager of the Rodmell Primary School, and have been for over twenty years President of the Rodmell Horticultural Society. It is extremely difficult to answer honestly the question why I have spent so many thousands of hours in these drab occupations. I do not really like sitting on committees and am not a good rank-and-file committee man, though I can be a very good secretary and even, when I take the trouble, a good chairman. There is, of course, a kind of childish or ignoble pleasure in the feeling of male importance which everyone feels when he takes his seat at a committee meeting. If you are chairman or secretary, you can feel at least a faint additional pleasure in the exercise of power, however feeble. Then too, as I have said, I find it extremely interesting to watch the psychological antics of five, ten, or fifteen men sitting round a table, each with his own selfish or unselfish axe to grind. I have always found the battle of brains and wills, boxing, wrestling, or ju-jitsu, with no blows or holds barred, which goes on round the table fascinating. Indeed, one of the reasons why I am ordinarily not a good rank-and-file committee man is that I tend to forget everything in the silent amusement of observing highly intelligent men fighting for Will-o'-the-wisps or even windmills as if for their own dear lives, converting their own pet molehills into God's Mount Sinais. Looking back over the aeons of slowly passing minutes that I have spent in the House of Commons and other less distinguished committee rooms, I must admit that I have enjoyed the spectacle of many great men or little men of great political expertise performing as if for my personal benefit; the wily, treacherous Ramsay MacDonald in the old I.L.P.; the mouselike Clem Attlee, who, when you least expected it, would suddenly show himself to be a masterful or even savage mouse, in the New Fabian Research Bureau; Bernard Shaw's gala performances of irrelevant wit and dialectic in the Fabian Society; the ruthless virtuosity of Sidney Webb in the Fabian Society; the new school of hard-headed, no non-sense, common sense of Harold Laski, G. D. H. Cole, Hugh Gaitskell, Hugh Dalton, Harold Wilson in the Fabian Society and Labour Party; the strange succession of Maynard Keynes, Kingsley Martin, John Freeman, and Jock Campbell on the *New Statesman* Board.

But these are, of course, not the main reason why I have continued for so many years to do all this dreary work. It had an object, a political or social object. My seven years in the Ceylon Civil Service turned me from an aesthetic into a political animal. The social and economic squalor in which thousands of Sinhalese and Tamil villagers lived horrified me; I saw close at hand the evils of imperialism and foresaw some of the difficulties and dangers which its inevitable liquidation would involve. When I returned to England after this seven-year interval, I was intensely interested in the political and social system; I could observe it with the fresh eye of a stranger, and also to some extent with the eye of an expert, for as Assistant Government Agent of a District, as a judge, and as a magistrate, I had learned a good deal about the art of government and administration.

My first contact with the economic system of capitalism in the England of 1912 was, as I described in *Beginning Again*, through a Care Committee of the Charity Organization Society in Hoxton. The immediate effect upon me I described in that book as follows:

One only had to spend a quarter of an hour sitting with Marny Vaughan on a Care Committee and another

153

quarter of an hour with the victim, Mr. and Mrs. Smith in the Hoxton slum, to see that in Hoxton one was confronted by some vast, dangerous fault in the social structure, some destructive disease in the social organism, which could not be touched by paternalism or charity or good works. Nothing but a social revolution, a major operation, could deal with it. I resigned from the Care Committee of the C.O.S. Hoxton turned me from a liberal into a socialist.*

A study of the co-operative societies of England and Scotland and seeing something of the lives of working class co-operators in the north confirmed my socialism. I became a member of the Fabian Society, the Labour Party, and the I.L.P.

The senseless war of 1914 deepened my conversion to a political animal. I was horrified by this spectacle of millions of human beings apparently driven by inexorable fate into communal madness, slaughtering one another by the million, scattering over the whole earth the most ghastly misery and pain and ruin, blindly destroying civilization in the name of civilization—and all this for objects which had no relevance, import, or even meaning for anyone outside a tiny ring of kings, rulers, aristocrats, statesmen, generals and admirals, and historians. I became obsessed by two questions: first, why human beings, and particularly Europeans, at intervals committed political and social suicide, like Gadarene swine, by rushing down a steep place into war. I could not accept the acquiescent resignation of the old Kaspars, little Peterkins, and little Wilhelmines contemplating the skulls, the

* *Beginning Again*, p. 100.

memorials of the Duke of Marlborough's victory at Blenheim or Lord Haig's "victory" at Passchendaele Ridge near Ypres:

> "But what they fought each other for
>   I could not well make out.
> But everybody said," quoth he,
>   "That 'twas a famous victory.
>
> They say it was a shocking sight
>   After the field was won;
> For many thousand bodies here
>   Lay rotting in the sun:
> But things like that, you know, must be
>   After a famous victory.
>
> And everybody praised the Duke
>   Who this great fight did win."
> "But what good came of it at last!"
>   Quoth little Peterkin:
> "Why that I cannot tell," said he,
>   "But 'twas a famous victory."

During the war itself I became absorbed in the problem of the fundamental causes of war and whether anything could be done to prevent it. During the last years of the war practically all my work was concentrated on this problem. The result was my book *International Government*, which originated from a report which I wrote for the Fabian Society. The deeper I went into the question, the more convinced I became that part of the solution depended upon the possibility of establishing some rudimentary form of international government. I did not believe that war could be abolished by international government, but came to the conclusion that war would almost certainly sooner or later be inevitable unless some sort of system of settling international disputes without war by methods of law or conciliation could be established. This led logically and practically to the idea of a League of Nations. In the Fabian Society, the Labour Party, and the League of Nations Society,* which I helped to establish, I worked with others to ensure that the creation of a League of Nations should be part of the peace settlement.

The League was created at Versailles. For the next twenty-seven years I worked as Secretary of the Labour Party Advisory Committee on International Affairs to try to get the Executive Committee of the Labour Party and secondarily the Parliamentary Party to make the League and the League system and its development the essence, the motive power, of their international policy. The Advisory Committee consisted of "experts" on foreign affairs, like Brailsford, C. R. Buxton, W. Arnold-Forster, Norman Angell, and Labour M.P.s who specialized in the same subject. As time went on, the scope of our work increased considerably; we "advised" the Executive Committee and through it the Parliamentary Party by a stream of reports and memoranda, explaining, often intellectually in words of one syllable, complicated situations and problems, warning about approaching crises, continually suggesting ways in which the Party's proclaimed general policy should be applied practically to these situations and problems. In the Fabian Society International Bureau I was doing the same kind of political work, but, whereas in the Labour Party Advisory Committee we

* See *Beginning Again*, pp. 191-2.

were trying to educate the Labour leaders, the political elite, in the Fabian Society we were addressing the rank and file as well as the elite.

As Secretary of the Labour Party Advisory Committee for Imperial Questions for twenty-seven years and in the Fabian Society Colonial Bureau I was trying to do the same kind of thing for imperialism, the Empire, the colonies which in the other committees I was trying to do for the League and international affairs. My aim, and I think the aim of the Advisory Committee and of the Fabian Society, was to put before the Labour Party and its rank-and-file supporters the facts and problems of the Empire and imperialism, to warn them of the dangers imminent in the inevitable demand for self-government and independence, to suggest a detailed, practical policy, varying from territory to territory, by which each should attain self-government or independence, and also by which economically and politically peoples could be prepared and educated for independence where that was not immediately possible.

I give these rather dreary political and institutional details because they are essential to finding a true answer to this question of the importance and effect of long years of "work", and of my "work" in particular. And I do not think that the particular aspect of the question, *my* work and *my* aims and *my* failure, is the only one involved; the unimportant particular case was related to and determined by the catastrophic historical events which led to the destruction of the League of Nations, Hitler's war, and the break-up of the British Empire. It is therefore in the light of this history that I ask myself the rather ludicrous question: What was the use of all this work? Was it

of the slightest importance? Did it achieve anything sub-
stantial of what it was intended to achieve?

Looking back at the age of eighty-eight over the fifty-
seven years of my political work in England, knowing what
I aimed at and the results, meditating on the history of
Britain and the world since 1914, I see clearly that I
achieved practically nothing. The world today and the
history of the human anthill during the last fifty-seven
years would be exactly the same as it is if I had played
pingpong instead of sitting on committees and writing
books and memoranda. I have therefore to make the
rather ignominious confession to myself and to anyone
who may read this book that I must have in a long life
ground through between 150,000 and 200,000 hours of
perfectly useless work. Objectively—I will deal with the
fact subjectively later—this is I think interesting, for it
throws some light upon the political determination of
history. There are thousands of people doing the kind of
political work which I did. The work has a clear, direct
object, to influence men's minds and so to alter the course
of historical events in one direction or another.

I was no fool at this particular game. It is not conceited
for me to say that in mind, temperament, and experience
I was peculiarly fitted for the kind of political work which
I tried to do. I have a clear mind, capable of quickly
understanding both theoretical and practical problems; I
proved in my seven years in Ceylon, by my rapid promo-
tion, that I was above the average in practising the art of
politics and government; I enjoy making difficult, dangerous, and "important" decisions and acting on them; until
age mellowed or emasculated me I suffered, as a politician,
from the disadvantage of regarding fools not with gladness

Malcolm Muggeridge
interviewing the author

Louie

but with exasperation and despair—and in politics the number of fools whom one has to suffer is terribly high—yet the many years in which I successfully managed the Labour Party Committees—difficult teams of intellectuals and trade unionists—proves, I think, that I did learn the art of managing and persuading all sorts and kinds of politicians; finally—and this is peculiarly important—by luck and the run of the game I very soon became known to and in many cases intimate with those people in the Labour movement who sat in the seats of power and who when the time came were Prime Ministers and Cabinet Ministers, Ramsay MacDonald, Clem Attlee, Sidney Webb, Hugh Gaitskell, Hugh Dalton, and many other worthy men. The relevance of what in the last sentence I believe, quite modestly, to be facts is that my failure to achieve anything was not due to personal political inadequacy and incompetence, and that, if I achieved nothing, it is almost certain that the enormous amount of similar work done by other people is and has been equally futile.

In order to explain and justify what I have just said, I will put down bleakly and objectively what seem to me to have been the positive and negative results of my 200,000 hours of labour. First the positive. I can, I think, chalk up one or two items of worldly success. I was interviewed for three days, eight hours a day, by Malcolm Muggeridge for a B.B.C. television programme. A sixty-minute T.V. interview by Malcolm on one's life and opinions is in some ways a popular apotheosis for someone like myself. Malcolm's power to confer the crown of notoriety upon the obscure is remarkable and I can give a significant proof of it. For many years once a year I have opened my garden to the public in aid of the Queen's Institute of District

Nursing. Up to and including April 1966 I never had more than one hundred visitors to the garden on the day it was opened. Malcolm's T.V. interview was in September 1966. The number of entrance paying visitors to my garden was 384 in 1967 and 457 in 1968. It is clear that Malcolm, by interviewing one, increases one's notoriety (or the notoriety of one's garden) by 284 per cent the first year and 357 per cent in the second.

My second worldly success was winning the W. H. Smith £1,000 award. But I am, of course, here concerned not so much with "success" in the wider worldly sense, as with the positive achievements, the positive effects of one's work and life. Well, I had some slight peripheral influence upon the establishment of the League of Nations and upon the constitution given to it. In *Beginning Again** I gave the facts which show that my book *International Government* "was used extensively by the government committee which produced the British proposals for a League of Nations laid before the Peace Conference, and also by the British delegation to the Versailles Conference". From 1920 to 1935 I worked incessantly through the Labour Party Advisory Committee and the Fabian Society, and also outside these organizations, to get British Governments, and of course Labour Governments pre-eminently, to strengthen the League, to use it as the main instrument of their international policy and of pacification and peace in Europe. This did have some effect. The Labour Party Executive and the Parliamentary Party did in fact adopt the policy which the Advisory Committee persistently recommended to them. With Will Arnold-Forster I prepared briefs for Arthur Hender-

* Page 189.

son to use—and he did use them—on the League; and
through Philip Noel-Baker I occasionally did something
of the same sort for Lord Cecil when he represented a
Conservative Government on the League Council.

I think the hundreds of hours which I spent working as
secretary of the two Labour Party Advisory Committees
—on international and imperial questions—had some
slight effects of a different kind. When after the 1914 war
the Labour Party began to recover from Lloyd George's
coupon election and became the alternative to a Conservative Government, when in fact the tide turned and at the
1924 election so many Labour M.P.s were elected that
Ramsay MacDonald formed the first Labour Government,
the class structure of the Party in the House gave it a
peculiar intellectual complexion. The great majority of
Labour M.P.s were working class and trade unionist, but
there was a small, influential minority—they held a dis-
proportionate number of Cabinet posts—of middle-class
intellectuals, many of whom had begun their political life
as Liberals. The trade unionists knew or thought they
knew everything that there was to know about the industrial and the economic system, but they were completely ignorant of and took little or no interest in foreign
affairs, the League, and the problems of empire. It is
significant that in 1924, when the future of the Empire,
colonies, and "colonialism" was one of the most important
of all political questions, MacDonald sent to the Colonial
Office as Secretary of State J. H. Thomas, an ignorant,
frivolous political buffoon. The Advisory Committees did
something to dispel this ignorance and apathy. Men of
knowledge and experience like Charles Buxton and Will
Arnold-Forster on the International Committee, and

Buxton and Sir John Maynard on the Imperial Committee, did an enormous amount of work between 1920 and 1930 to educate the Party. A certain number of Labour M.P.s joined the Committees and regularly came to the meetings. We continually briefed them for debates in the House, and gradually we helped to create a nucleus of M.P.s with a real knowledge and understanding of the problems. M.P.s like Sir Drummond Shields, who became Parliamentary Under-Secretary of State, India Office, and later the Colonial Office, and Arthur Creech Jones, who became Secretary of State for the Colonies, always acknowledged that their education in imperial politics came very largely from the Committee. We also provided the Party with an "advanced" policy with regard to the League of Nations, India, and the Empire or, as it became, the Commonwealth. When Charles Buxton and I in the early 1920's produced a detailed programme for developing and educating the African colonies for self-government and the Labour Party adopted and published it as their official policy, it seemed as if we had really accomplished something important.

It was a delusion. The work which we were doing was, of course, closely connected with the most historically important events between 1919 and 1939 and it created for us the kind of magnetic field which I have described above. It had the usual effect of such magnetic fields on all of us. However often one has been disillusioned one almost inevitably feels some (uneasy or even perhaps guilty) sense of importance when one goes down to the room of the Prime Minister, even if it is only Ramsay MacDonald, in the House of Commons to discuss with him a difficult and dangerous international crisis or the next tottering steps of India to self-government. The whole thing was as phoney as the Prime Minister. Take for instance the case of the League of Nations. I still believe that after the 1914 war the only hope of preventing a second war was in the establishment of an effective League, the beginnings of a new international order, based upon law, collective security, and the pacific settlement of disputes. I believe too that if British Governments had gone all out for establishing the League and had used it as the instrument of their policy in the turmoil of peace which followed inevitably on the turmoil of war, they might have succeeded, they might have obtained sufficient support from the other states of the world to persuade or even compel first victorious France and later a renascent Germany to work for peace instead of pursuing policies which could only end in war.

There was in fact never any real hope that this would happen. Conservative Governments and statesmen, from Baldwin and Samuel Hoare to Neville Chamberlain, never believed in the League or attempted to use it; though they occasionally paid lip-service to it, they thought that they themselves were "practical men" and that the League was a gimmick of "idealists". It is the practical men, not the idealists, who ever since the dawn of history have, by their practical policies, produced the unending series of disasters, the catalogue of miseries, which we call human history. They produced Hitler and the second great war. The only Conservative statesman who saw that the collective security system could be used to prevent that war was Churchill—and it was too late when he was at last converted. As for the Labour statesmen and the Labour Governments, to work with them in the 1920's was a lesson in frustration. The Governments were shortlived and without a majority in the House. Ramsay MacDonald shilly-shallied in foreign policy as he did in other things. Only Henderson at the top had a real grasp of what a League policy meant and he was not supported by Ramsay. When, as I recorded in *Downhill All the Way,** we urged the Labour leaders to offer to join Winston Churchill in a coalition Government as the last chance of deterring Hitler from starting a world war—the very step they took when it was too late and Hitler had started his war—not one of them would even consider it seriously. Statesmen, those who are supposed and pretend to control events, are almost always content complacently to be controlled by them. That is why, while scientists produce bombs so efficient that they could destroy the whole human race in the space of half an hour and can send men to the moon and back, statesmen and governments allow international relations and the peoples of Europe to be controlled for years by a psychopathic lunatic and tolerate political and economic chaos from Vietnam to Nigeria and from Moscow to Washington.

My work with regard to the Empire and imperialism was just as futile. In a way it was even more exasperating. During the 1920's there were two problems of primary importance: first, to work out with the inhabitants of territories like India, Burma, and Ceylon the methods by which they could pass immediately from subordinate status to independence; secondly, to prepare those territories, mainly African, not ripe for immediate independence, by education and economic development so that they could pass, as rapidly as possible, through stages of

* Page 246.

self-government to political independence. When we put
this before the Labour leaders and bigwigs and worked
out in some detail the process by which the policy could
be implemented, our proposals, as I have said, were
accepted and put out as the official policy of the Party.
When the time came for Labour Ministers and Governments to put their policy into practice, they almost always
failed to do so. In my memory two incidents (which I have
previously described in *Downhill All the Way**) stand out
as characteristic of Labour Ministers. In the first Charles
Buxton and I, as Chairman and Secretary of the Advisory
Committee, sat fantastically in the empty House of Lords
one on each side of Sidney Webb, then Lord Passfield and
Secretary of State for the Colonies, and vainly urged him
to carry out the Party's promises and insist that some
nugatory sum should be included in the Kenya budget
for the education of African children and the provision of
roads in the African Reserves. In the second, once more
with Charles Buxton, I sit in No. 11 Downing Street on
one side of a long table and on the other side sits Clem
Attlee, not yet a peer, but Deputy Prime Minister and
Lord President of the Council in Churchill's war Cabinet; we met a stony refusal from him when we urged him
in a recurrent crisis in Indian affairs to do everything
possible on the lines of the Labour Party's declared policy.

Of course I may be completely deluded in thinking that
the policies of the League in international affairs and that
rapid progress towards independence and self-government in imperial affairs might have saved the world from
Hitler and the war and might have ensured a less bloody
and chaotic break-up of empires. There are however some

* Pages 237 and 228.

facts which make it more than possible that I am right. No one can deny that the policies actually pursued have produced in the last thirty years more horrors, misery, and barbarism than occurred in any other thirty years of recorded history. It is surely significant that over and over again the measures which we were urging in the 1920's and were rejected by "practical" statesmen as utopian were adopted by them some twenty or forty years later— adopted when it was too late for them to be effective. After all, practical statesmen, who had refused to use the League of Nations before the war as idealist and utopian, resuscitated it under the name of the United Nations after the war and are misusing it in exactly the same way as they did its predecessor. I have given in *Downhill All the Way** reasons for believing that "the perpetual tragedy of history is that things are perpetually being done ten or twenty years too late", and I will not repeat them here.

As to the subjective effect of looking back over a life of eighty-eight years and 200,000 hours of work and of coming to the sobering conclusion that they have been, if not completely futile, at least mainly ineffective, there are two aspects of this picture. First there is the direct effect upon me of the state of the world, the climate of civilization or barbarism. I feel passionately for what I call civilized life; I hate passionately what I call barbarism. When as a small child I heard my father say one day at lunch that, as regards rules of life, a man need only follow that advice of the prophet Micah: "What doth the Lord require of thee, but to do justly, and to love mercy, and to walk humbly with thy God", I am sure that I did not really understand what he was talking about, yet in some

* Page 225.

In Israel: Trekkie and the author by the Sea of Galilee

curious way, I think, the words entered into and had a profound effect upon my mind and upon my soul, if I can be said to have a soul—for though it must have been more than eighty years ago, I can still see the scene, all of us children sitting round the Sunday lunch-table, the great sirloin appearing from under the enormous silver cover, my father with his serious, sensitive face with the carving knife poised over the sirloin as he quoted the prophet Micah, and the rather surprised and sheepish face of my cousin Benny who was not prone to walk or talk humbly with his God or anyone else.

Like Benny, I have never been much concerned with God or with walking humbly with him, but I believe profoundly in the other two rules. Justice and mercy—they seem to me the foundation of all civilized life and society, if you include under mercy toleration. This is, of course, the Semitic vision, but, when later I found that the Greeks had added to it the vision of liberty and beauty—τὸ καλόν καὶ ἀγαθὸν—I saw, when I added the words of Micah to the speech of Pericles in Thucydides, what has remained until today my vision of civilization. And my feelings with regard to communal justice and mercy and toleration and liberty are both ethical and aesthetic, and it is this combination which gives to my feeling about what I call civilization both its intensity and also a kind of austerity. The visions of civilization and the partial, hesitating, fluctuating activation of these visions in the barbarous history of man, and the classical instances in which individuals have risked everything in a fight for justice, mercy, toleration, and liberty against the entrenched forces of kings and emperors, states and establishments, principalities and powers, all these have always given me not only an intense feeling about what is good and bad, what is right and wrong, but also the kind of emotion which I get still more powerfully from a play of Sophocles or Shakespeare, the Parthenon or the Acropolis, a picture of Piero della Francesca, a cello suite of Bach or the last movement of the last piano sonata of Beethoven. Actual examples are the description of Athenian civilization by Pericles; the process of abolishing the slave-trade and Pitt's speech as the sun rose upon the debate in parliament which had lasted all night; Voltaire in the Calas case and Zola in the Dreyfus case; the passionate campaigns of Gladstone for liberty in Ireland and for justice and mercy in Armenia.

All this is the positive aspect of my political philosophy, if that is not too pretentious a name for my political beliefs, feelings, desires. The negative aspect is more relevant to the feelings with which autobiographically I look back on the effect of my 200,000 hours of work. Injustice, cruelty, intolerance, tyranny fill me with a passion of anger and disgust, and again my feelings are both aesthetic and ethical. To watch the Governments of Britain, France, and the United States destroy the League as a potential instrument of peace and civilization against Hitler and the Nazis; to see the savage insanity of Hitler ecstatically infecting millions of Germans and carrying Europe and the world inexorably downhill into war; to observe the crude cruelty and stupidity of Soviet communism and the Iron Curtain from Stalin's massacres to the invasion of Czechoslovakia; to watch millions of ignorant Chinese and many western Europeans, who should know better, hail the imbecilities and savagery of Mao and Chinese communism as divine political and economic revelations; to see Americans at one and the same time showing superhuman skill and intelligence by sending three astronauts round the moon and year after year fighting in Vietnam a stupid, unjustified, bloody, and useless war; to see, what one had hoped for, the break-up of imperialism and colonialism, and then to find in the place of empires the chaotic crudities and hydrogen bombs of Mao in China, the senseless hostility of Pakistan and India, the unending war of Arabs and Israelis, the primitive brutality of apartheid in South Africa and Rhodesia; the ebb and flow of chaos and bloodshed and bleak authoritarianism in the new independent African states—when I look round at these facts in the world of today, I feel acute pain, compounded, I think, of disappointment and horror and discomfort and disgust.

Such are my reactions to the facts; except in the word "disappointment" they throw no light upon my reactions to my own failure and futility. I do not think that ultimately I am much concerned by failure or success as a subjective experience. When I was secretary of the Labour Party Advisory Committees irate members used from time to time to come to me and say that we should give it up; we did, they said, an enormous amount of work for the Executive and the Parliamentary Party, and nothing came of it; the Party accepted our memoranda and reports and even adopted our policies—and yet nothing really came of it. I tried and usually succeeded in soothing them. I pointed out that we did occasionally get some important policy officially adopted and even at rare intervals acted upon; we did help the Executive and M.P.s to understand the crucial international and imperial problems; we occasionally prevented some

unimportant or even important politician from going hopelessly astray. It was a question of casting bread on the
waters—Heaven rejoicing in the conversion of a single
sinner—being contented with small mercies or achievements—etc., etc.

These crumbs of comfort with the slightly Pecksniffian
or Micawberish clichés to sweeten them seemed, as I
said, to soothe and even convince the M.P.s and other
frustrated members of the Committees. I do not think that
they ever convinced me or were the real arguments or
motives which kept me doing for so long so much work
for such invisible and probably imaginary results. My
attitude to these things was and, I suppose, still is rather
different. For the vast majority of men, who are not great
men or great criminals or both, there are three alternative
ways of resigning oneself to one's own impotence to control one's own fate or to affect in any way the march of
events: the first, which the enormous majority of human
beings adopt from birth to death, is to ignore it all, make
the best of it, earn one's living, marry a wife or a husband,
join a club, play a game of golf, do the pools, watch the
telly, eat and drink for tomorrow we die, and in fact
finally die. The second method of ordering one's life in
what for them is a hostile world and a hostile universe is
adopted by a small number of extremely intelligent and
sensitive people. Most of them are artists—writers,
painters, or composers—if they are great, they create a
new world both for themselves and for other people. But
I have known a very few people in this class who are not
artists or creative who have found a slightly different way
of defying fate. They are defeatists; they give up the
struggle; they spin for themselves, and live in a cocoon

of unreality and eccentricity. The most curious example of this kind of cocoon existence was that of Saxon Sydney-Turner whose character and way of life I tried to describe in the first volume of my autobiography, *Sowing*.*

The third method is one which, with many other people, not by conscious intent so much as through innate character, I have myself pursued. I cannot disengage myself from the real world; I cannot completely resign myself to fate; somewhere in the pit of my stomach there is a spark of fire or heat which at any moment may burst into flame and compel me violently to follow some path or pursue some object—no doubt too often some shadow of a dream or political *ignis fatuus*—contrary to the calculations of reason or possibility. It is in the pit of my stomach as well as in the cooler regions of my brain that I feel and think about what I see happening in the human ant-heap around me, the historical and political events which seem to me to make the difference between a good life and a bad, between civilization and barbarism. I have no doubt that, if at any moment I had become convinced that my political work produced absolutely no effect at all in any direction, I would have stopped it altogether and have retired to cultivate my garden—the last refuge of disillusion. But the shadow of the shadow of a dream is a good enough carrot to keep the human donkey going through three score years and ten (and in my case even four score years and eight), and that is why I never could and still cannot yield even to the logic of events. On these things apparently I will "not cease from mortal fight" and my sword will not "sleep in my hand", although I know quite well that not Jerusalem, but only

* Pages 103-8 and 113-9.

hideous red brick villas will be built in what was once "England's green and pleasant land".

All these excuses and explanations of why I have performed 200,000 hours of useless work are no doubt merely another way of confessing that the magnetic field of my own occupations produced the usual self-deception, the belief that they were important. That brings me to the final excuse or explanation. As I have said before, all through my life I have always believed and, I think, acted on the belief that there are two levels or grades of importance. *Sub specie aeternitatis*, in the eye of God or rather of the universe, nothing human is of the slightest importance; but in one's own personal life, in terms of humanity and human history and human society, certain things are of immense importance: human relations, happiness, truth, beauty or art, justice and mercy. That is why in his private and personal life a wise man would never take arms against a sea of troubles; he would suffer the slings and arrows of outrageous fortune, saying to himself: "These things are momentarily of terrible importance and yet tomorrow and eternally of no importance". And in a wider context, though all that I tried to do politically was completely futile and ineffective and unimportant, for me personally it was right and important that I should do it, even though at the back of my mind I was well aware that it was ineffective and unimportant. To say this is to say that I agree with what Montaigne, the first civilized modern man, says somewhere: "It is not the arrival, it is the journey which matters".

Chapter Four

ALL OUR YESTERDAYS

Looking back over one's life, one of the curious things one notices is how two or three small events happened years and years ago, then for years the consequences disappear beneath the surface of one's life like underground springs or streams, and then, like streams breaking out of the ground as tributaries to form a great river, years later the events reappeared with important consequences in one's life. This kind of thing happened to me in 1943. It began with a letter from Phil Noel-Baker nearly twenty years before in the early days of The Hogarth Press. Phil was for a time in the Secretariat of the League of Nations in Geneva. I got a letter from him asking me whether I could find a job for a young woman, Alice Ritchie, who had just been sacked, he thought rather unfairly, from a responsible post in the Secretariat. Alice was a remarkable young woman with a character entirely her own. I cannot now remember what brought about her downfall, but I think it was some inexcusable criticism of a superior which the superior ought to have excused.

I saw Alice and offered her the post of traveller for The Hogarth Press. She was the first woman to become what is called a publisher's representative in Britain. She proved to be a good, if unconventional, commercial traveller, and on the whole she enjoyed taking The Hogarth Press books round the booksellers and getting orders from them. But she was much more than a traveller in books. She also wrote them. She was born in South Africa and had been educated at Newnham, but her roots were in the north of Great Britain for her father, an architect, had been born in Scotland and her mother in Durham. She had a brother, Pat, and a younger sister, Trekkie. The whole family returned to England for good during the 1914 war, because her father, who had already fought in the Boer war, decided that he must fight for Britain in the first great war. Pat, who was still in his teens and had a passion for flying, joined the R.F.C. He remained in the R.A.F. after the war and reached the rank of Air Vice-Marshal. I look back with some amusement to his attempts in the early 1920's to prove to me that flying was already a reliable form of transport. I had rather tactlessly said to Alice that, in my opinion, it was not yet reliable. Pat was at that time stationed in the West Country but once a week had to come to London to attend the Air Ministry. When Alice reported to him what I had said, he offered to take me up in a plane from Hendon and prove I was wrong. For many weeks after this I was rung up by him every Friday, and every Friday the weather was not good enough for a flight. However, at last one day he rang me up to say that it was all right and Virginia and I set off to Hendon. Pat met us and took us to the airfield. There was a long wait and then the disconsolate Pat came and told us that owing to some hitch the plane could not be used. Next week he was moved to another station and to my regret I never left the earth in a plane piloted by Pat Ritchie.

Alice was, as I said, a remarkable young woman and too good to spend much of her life travelling books. She wrote two novels, *The Peacemakers* and *Occupied Territory*, which we published in The Hogarth Press. The first was based on her experience in the League of Nations, the second on her experiences in occupied Germany after the war, when her father, who had reached the rank of colonel, was in command of his regiment there. Both these books showed real talent and considerable promise. She had the mind of a novelist and the temperament of an artist, but there was a psychological twist in her which made it almost impossible for her to face the final word or stroke or note in artistic production. She belonged to that strange race of Penelope writers who unpick every evening all the stitches which they sewed in the morning. Day after day Alice would write all the morning and tear it all up in the afternoon. This disease of artificial sterility is not uncommon among writers, particularly novelists, and some who suffer from it are potentially very good writers. The cause of the disease is not always the same. In most cases, I think, it comes from a horror of cutting the umbilical cord which binds the work of art to the artist, a refusal, usually unconscious, to throw the child to the wolves, the book to the reviewers or critics. Virginia, as I have said before, had this horror of the cutting of the umbilical cord, but, like several others of her family, she combined nervous instability and skinless hypersensibility with remarkable mental toughness, and the moment always came with every book she wrote when she said: "Publish and be damned".

Alice, I think, was not without this umbilical horror, but the principal cause of her artistic sterility, of her inability to finish a book, was different. It was the same psychology as Desmond MacCarthy's. "The best is the enemy of the good," said the Greeks. Desmond's standards as an artist were so high that, knowing that he could never in practice attain them, he never really even attempted to write the novel which he so often talked of writing. There was in this shrinking and shirking a mixture of mental laziness and artistic cowardice. Alice suffered from the same kind of inhibitions. She was not as mentally lazy as Desmond, but she had the same fear of committing herself artistically. She did not, as he did, invent good and bad reasons for not writing at all; she wrote her novel daily before midday and tore it up before midnight. In the hope of curing her of this disease, at one time I got her to post off to me every morning what she had written as soon as she had written it, but this did no good. After a week or more she got bogged down and I had to send her back what she had sent me. Penelope started all over again, and her third novel was never written.

There was a streak of melancholy in Alice which may have been part cause or part effect of her sterility as a writer. After some years she gave up her job with The Hogarth Press, but Virginia and I liked her very much and continued to see her from time to time. But in the 1930's she disappeared into Palestine, for she went to keep house in Jerusalem for her brother Pat who had been seconded and lent by the R.A.F. to the Army in those troubled years. I heard little or nothing from her and she seemed, as sometimes people do, to have passed out of my life. But in the middle of 1941 I received a letter from her asking me whether I would come and see her in London as she was ill.

Pat, who was now a Group Captain, had been brought back from Palestine for war duties in England, and Alice had returned with him. I found her in Victoria Square,

already very ill, dying of cancer. For the next few months
—until, in fact, the cancer killed her—whenever I was in
London I used to go and see her. She seemed to be
facing death bravely, with her eyes open, and we talked
on the surface as we had always talked. Yet we talked
under the shadow of death. When I see someone, particularly someone who like Alice is young, dying, when I sit
talking to her knowing that she is dying, and in her case
she knowing that I knew that she was dying—I get a
terrible feeling that time has stopped, the earth is no
longer revolving, the universe has slowed down—we are
waiting in a void for the final catastrophe, the passing of
life.

I do not think that there is sentimentality or self-
deception in what I have just written. However sophisticated and atheistic a man may be, there is in his bowels,
in the pit of his stomach, at the bottom of his heart, in the
convolutions of his brain, a primeval attitude towards
death. I myself feel it and have always felt it strongly. I
saw the puppy dying in the bucket of water; in Ceylon at
the Pearl Fishery, with the dawn coming up like thunder,
I looked down on the dead Arab lying on the sand at the
sea edge*; I watched Alice dying of cancer in Victoria
Square; years later, only a day before he died, I went to
see Clive Bell, dying of cancer, no longer able to talk, but,
in the room in which time had stopped, his eyes watching
for death, but still eager to hear from me the trivialities of
living; I saw Virginia's body in the Newhaven mortuary.
In all these cases, however acutely my own personal feelings were involved or not involved, I had the primeval
sense of time stopping, the universe hesitating, waiting,

* See *Growing*, p. 95.

in fear, regret, pity, for the annihilation or snuffing out of a life, of a living being.

If you are the Pope or the Archbishop of Canterbury or the rector of Rodmell or the Queen of England or the headmaster of Eton or the Lord Chancellor or the Director-General of the B.B.C. or a less distinguished person sufficiently terrified by the prospect of dying, you believe that absolute truth about life and death and immortality was revealed three, four, or five thousand years ago to Semitic savages in the sands of Sinai and in the craggy town of Jerusalem—or if you cannot quite swallow that, you believe that absolute truth about immortality can be found in the pit of one's stomach where by some strange psychological alchemy primeval fear and wishful thinking produce absolute truth. I wish I could believe this, but I have no faith in the pit of my or anyone else's stomach, and I see no more reason for accepting the dreams and nightmares of Semitic savages on death and immortality than on the nature of thunder or the origin of species.

Alice died. Some time later I had a letter from her sister Trekkie; she said that she had been ill, but was now practically recovered, and she asked me to come and see her. I had seen her once or twice in the far-off days when Alice was with us in the Press. She was then a very young woman, a painter at the Slade, extremely beautiful. She designed the jackets for Alice's novels and she did a few other jackets for us at that time. I do not think that we met at all between 1930 and 1942. She had married Ian Parsons, the publisher, a Director of Chatto & Windus, and they lived in a house in Victoria Square—Alice in the last months of her life was living with them.

I went to see Trekkie and I dined with her and Ian in Victoria Square, but I did not see much of them until towards the end of 1942. It was one of the gloomiest periods of a gloomy war. Ian was in the R.A.F., in the bowels of the earth in Westminster, poring over photographs of airfields in occupied Europe. Trekkie every day rode on a motor bicycle to a super-secret department of the War Office in Petersham. But towards the end of the year I began to see more of them and Trekkie came and stayed for a weekend at Rodmell. This altered both the rhythm of my life and the future of The Hogarth Press. As regards the Press, it was because I had got to know Ian so well that, as I have described in a previous chapter, when John Lehmann put the pistol of dissolving our partnership at my head, I went straight off to Ian and asked him whether he and his partners in Chatto & Windus would step into the shoes of John in The Hogarth Press. Hence the fortunes of the Press (which John, quite correctly, accused of practising amateur publishing and which was soon to be transformed into a more dignified Limited Company) and my amateurish career as a publisher were in the next twenty-two years indissolubly, happily, and profitably linked to those of Ian and Chatto.

As regards the rhythm of my life, as 1943 waxed and waned, I began to see more and more of Trekkie both in London in the dilapidations of Mecklenburgh Square and at weekends from time to time in Rodmell. In October I moved from Mecklenburgh to Victoria Square and so, when in London, I was living next door to the Parsons. In 1944, after the invasion, Ian went to France and Trekkie came and stayed at Monks House. They gave up their house in Victoria Square and decided that after the war they would have a house near me in Sussex. We searched the neighbourhood for many months and they eventually found and leased a house in Iford, which is the next village two miles away from Rodmell. We also arranged that they should share with me the first and second floors of my house in Victoria Square.

Thus when at last the war ended and Ian was demobilized, a new rhythm of life began for us. I stayed one or two nights a week in Victoria Square to do my work at The Hogarth Press and various political committees; Trekkie did the same, but stayed with me at Rodmell in the middle of the week. We cultivated our gardens passionately, the Parsons at Iford and I at Rodmell.

In 1946, when this rhythm of life began for me, I was sixty-six years old, a time of life when one has passed or is passing from middle age into old age, when in most spheres of life one is officially called upon to retire on a pension. I have slowly in the last twenty-two years shuffled off most of my political activities. In 1946, having done twenty-seven years of hard labour, I resigned from the Labour Party Committee and was slightly amused when I was informed by the Secretary of the Party that the National Executive had passed "with acclamation" a resolution "placing on record its deep appreciation of your great services to the Party by enthusiastic and persistent hard work, through many years and over periods of great discouragement". I think that many of my fellow-members would have said that a good deal of the discouragement had come from the National Executive itself. However, the wise man gratefully accepts the fact that, if he is ever given a bouquet, it will only be when he resigns or dies.

In Ceylon: Mr and Mrs Fernando, the author, and Trekkie

In Ceylon: the author on a catamaran

In 1967 The Hogarth Press celebrated its fiftieth birthday. It is rather surprising that it should have existed for half a century and that it was still flourishing fifty years after Virginia and I began printing *Two Stories* by Leonard and Virginia Woolf in the dining-room of Hogarth House in Richmond.* I was thirty-seven when we printed and published our first book and I was therefore eighty-seven when the Press completed its half-century. It was and is, I suppose, equally surprising that I was and am still a Director taking an active, if modified, part in its activities. The advantage—perhaps, too, sometimes the disadvantage—of being what is called by the Inland Revenue self-employed is that there is no one who can authoritively ask one to resign. Being in a modified degree still able to walk, see, hear, and think, I can still go up to London once a week to the office, read MSS, and deal with the larger problems of publishing.

On the surface The Hogarth Press, as a publishing business, has changed its nature considerably since John Lehmann left it in 1946. It is a limited company and I have several co-directors. In fact, though in the business and technique of printing, binding, selling, distributing, and accounting it is now part of a large and efficient organization very different from the business machine which we relied on during the first thirty years of its existence, in its spiritual or intellectual nature it has not materially changed.

It is, of course, one of several subsidiary companies of Chatto & Windus Limited. When Ian in 1946 agreed that Chatto & Windus should step into John's shoes, the Directors of Chatto were, as I have said, Harold

* See *Beginning Again*, pp. 231-7.

Raymond, Ian, Norah Smallwood and Piers Raymond. The attitude of all four towards books, literature, and publishing was the same as mine and consequently the list of books published by Chatto was fundamentally merely a very much larger version of the list of books published by The Hogarth Press. That is why we have never had the slightest disagreement as to what The Hogarth Press should be and do.

I think that our publications, therefore, during the last twenty-two years have not changed in number, nature, or standard. The Press remains a small publishing business with an annual list of at most ten or twelve books. We are and have always been unashamedly highbrow publishers, but in the last twenty-two years, as in the previous thirty, we have had a surprisingly large number of best-selling and long-selling books and authors.

In the previous paragraphs I have been dealing with the effect of the rhythm of my life since the war upon my political and publishing occupations as old age crept upon and over me. I have gradually disencumbered myself from many of my previous responsibilities and I now spend much less time on them than I did twenty years ago. The third great occupation of my life has been writing, and as I have ceased to be a politician and a publisher, I have become a much more prolific author. I published *Principia Politica* in 1953, *Sowing* in 1960, *Growing* in 1961, *Beginning Again* in 1964, and *Downhill All the Way* in 1967. I am, I think, a person who habitually throughout his life has got a great deal of pleasure from a great variety of things. Eating and drinking, reading, walking and riding, cultivating a garden, games of every kind, animals of every kind, conversation, pictures, music,

friendship, love, people—all these things give almost un-
failingly pleasure, varying, no doubt, in quality and intensity. The pleasure in all these things is not only very
pleasant but also, I think, very good; indeed, the only
pleasures which are bad are pleasures in other people's
pain, aggressive and sadistic pleasures, pleasure in what
is on the dark side of the moon. The most civilized
civilizations have always counted pleasure to be a very
good thing, and the most uncivilized civilizations have
always puritanically frowned on happiness. I do not find
that old age has decreased my capacity for pleasure,
though it has, of course, destroyed one's ability to do some
of the things which used to give one pleasure. It is regret-
table that the impotence of the aged body results in my
inability any longer to play cricket, football, squash
rackets, hockey, or lawn tennis, and to function in still
more important ways. But, if I can no longer have the
satisfaction of hitting a six over long-on's head, I can still
play a good game of bowls, and it is possible to find great
happiness in love and affection long after one has to
accept the fact that all passion is spent. And one of the
pleasures which all my life I have found to be most reliable
and to have remained unaffected by the vampirism of
senility is the pleasure of writing. It is, oddly enough, a
physical as well as a mental satisfaction. I like to feel the
process of composition in my brain, to feel the mind
working, the thoughts arranging themselves in words, the
words appearing on the virgin white paper. When one is
actually writing, one is concerned only with the process of
thought and composition; one is not concerned at the
moment with whether the result is good or bad—though
that awkward question will have to be put and answered

later. Sufficient for the morning is the pleasure thereof, and one of the most unfailing pleasures is to sit down in the morning and write.

Another pleasure of the last twenty years which has not been much dimmed by age is travel. Travellers' tales autobiographically are almost always boring, and I must avoid them, but there are one or two of my post-war journeys of which I must say something. One of the many deprivations of war that I have felt acutely is the fact that one is cut off from the rest of the world. A great deal of sedentary sediment seems to be rinsed out of the mind as soon as one has crossed the Channel and one sees again, if one is going to drive south, the straight white road of France which will take you to the Mediterranean. From 1939 to 1949 I never saw any sky but that which is bounded on the south by the very English Channel. It was a tremendous pleasure in 1949 once more to travel along French roads and I have done it a good many times since 1949.

I think that one of the great pleasures which I get from travel is the casual meeting and talking with some stranger in the foreign land with whom one is instantly in sympathy and understanding. It is not only their personal relationship which, for a moment, can make one forget the inveterate savagery, the ingrained nastiness of human beings; there is also the pleasure of suddenly getting a glimpse of a civilization and barbarism widely different from one's own. I recall two instances of this. The first was in Greece on the Acropolis. I was sitting on the parapet overlooking the Agora when a man, the tout selling postcards, came up to me and said: "Where did you get that walking stick, Sir?" I said that I had bought it the week before in Sparta. "It is not a Greek stick," he said;

"and it is painted; it is not a very good stick." I said that I knew that, but I had a habit of buying a stick in any foreign country I visited. "It is not a very good stick," I said, "but it is a rather curious stick, and, whenever I use it in England, I shall remember the market-place in Sparta and Greece and the Acropolis and you." He laughed and sat down by my side and began to talk of Greece and England. An hour later we were still sitting on the parapet of the Acropolis talking. We began by talking about Greece and the economic conditions and Greek politics, and we ended by talking of life, his life and mine. At last he got up, took off his hat, and shook my hand, and said: "Thank you, Sir; I have much enjoyed our talk." I thanked him and said that I had much enjoyed our conversation. The Greeks, I think, are politically among the most un-civilized people of Europe, but in other respects they are the most civilized and intelligent. The intelligence, know-ledge, humanity of this man were extraordinary. I do not think that there is any other country in the world, except perhaps Israel, in which it would be possible to have the kind of talk and relationship which I had with a tout sell-ing photographs.

The other incident happened in Israel, to which I travelled with Trekkie in 1957. In the 1920's I was against Zionism and neither the great Namier nor the still greater Chaim Weizmann could convert me, though they tried to do so. I think that the whole of history shows that the savage xenophobia of human beings is so great that the introduction into any populated country of a large racial, economic, religious, or cultured minority always leads to hatred, violence, and political and social disaster. Whether it is the Negroes in America, white settlers in

Kenya and Rhodesia, or West Indians in Wolverhampton the miserable story is always the same, and, therefore, until the human race becomes more civilized everything should be done to prevent the creation of new centres of conflict between minorities and majorities. That was why I thought originally that the Balfour Declaration and the introduction of a Jewish minority and a Jewish state in a country inhabited by a large Arab population was politically dangerous. But in politics and history once something has been done radically to change the past into a new present, you must act not upon a situation which no longer exists, but upon the facts that face one. When the Jewish National Home and hundreds of thousands of Jews had been established in Palestine, when Hitler was killing millions of Jews in Europe, when the independent sovereign state of Israel had been created, when the Arabs proclaimed their intention of destroying Israel and the Israelis, Zionism and anti-Zionism had become irrelevant. At any rate I went to Israel, with a comparatively open mind, to see for myself this return of the Jews to Jerusalem.

I have never felt so exhilarated—not even in Greece—by the physical climate of a country and the mental climate of its inhabitants. In Ceylon in the low country I lived for several years in a climate which has no autumn or winter and only an apology for spring; it is perpetual summer, a very hot, dry, and sometimes oppressive summer at that. I have not yet discovered any climate too hot for me, for I would always rather be too hot than too cold, and since most of my life I have lived in England, for most of my life I have been much too cold. I liked the heat of Jaffna and Hambantota in Ceylon, but sometimes it lay upon one like a physical weight; it was as if one was in bed with too many blankets on one. In Jerusalem and Haifa and Safed it was hot enough even for me, but there was an extraordinary sparkle and freshness, a keenness without a touch of cold in it. This was immensely exhilarating, but what was even more exciting was that the same qualities existed in the inhabitants. The physical and mental effervescence in the streets of Tel Aviv is something unlike anything I have ever felt in any other country. It reminded me of the busy buzz of productive ecstasy on the running-board of a hive on a perfect summer day and hundreds of happy bees stream in and out of the hive on the communal business of finding nectar and storing honey.

I had some introductions and therefore got to know a certain number of people, intellectuals all of them, journalists, writers, teachers in the university and in schools. But I also was able to see something of the kibbutzim and their inhabitants; moreover, Israel is a country in which the inhabitants everywhere talk to the stranger. What astonished me was the immense energy, friendliness, and intelligence of these people. It was the vision of a civilized community creating materially out of the rocky soil and spiritually out of the terrible history of all the peoples of the world, in all the millennia since Adam, a new civilized way of life. There was another curious thing in the mental atmosphere of Israel. It reminded me of the atmosphere of London during the worst days of the blitz. There was a feeling everywhere of comradeship, solidarity, good-will, and I suppose this was due, as in London, to the sense of a common menace and danger. It is impossible anywhere in Israel ever to forget that it is a small country surrounded by a ring of violently hostile Arab states, and people who consider themselves to be perpetually at war with, and pledged to destroy, Israel. It is, I suppose, only the threat of this kind of imminent death or of German bombs raining down upon one from the skies which can make the human herd feel its own unity and humanity.

The particular incident which gave me a glimpse into this civilization of the Israelis occurred in Tiberias. We took a taxi to Nazareth. The taxi-driver began to talk to us as soon as we left Tiberias. Were we Roman Catholics? No. Were we Church of England? No, we were agnostics or atheists, though I myself had been born into the Jewish religion. He was, he said, glad to hear it, for now he could tell us the truth about the places through which he would be driving us. "I do not like to upset people," he said, "and in many cases I find that it is impossible to tell the truth about what has happened in this part of the country without upsetting Roman Catholics and Protestants." He was a remarkable man. He had an intimate knowledge of the country between Tiberias and Nazareth and a lively interest in and love for it. As we drove along he gave us a running commentary on the ancient and contemporary history of almost every small town, village, and hamlet through which we passed. His knowledge of world history was far greater than Trekkie's or mine. He spoke perfect English and I have never heard a talk or lecture in which knowledge, imagination, and humour were more happily blended. He was a man of quiet gentleness, humanity, and good-will. One felt the intensity of these qualities in him the moment one saw him talking with other people. The land between Tiberias and Nazareth is largely inhabited and cultivated by Arabs who did not leave Israel, as so many Arabs in other parts did, during the Arab-Israel war. Our driver several times stopped our car and took us to see some small mosque or other item of interest in an Arab village. It was obvious that he was well known and very well liked everywhere. Nazareth is a mainly Arab town and, as we drove into it, he said that he would hand us over to an Arab to show us the sights. "I do not think it is fair," he said, "for a Jew to show visitors over an Arab town."

There is no other country in the world—except, perhaps, Greece—where you could pick up a taxi in the street and find that it was being driven by someone with the intelligence, knowledge, and humanity of this man. To find such a civilized taxi-driver in Israel, however, was not, I think, quite as remarkable as finding that very civilized seller of postcards in Athens. The driver came from a highly educated and sophisticated middle-class background, for his father had been a Professor in a Jugoslav university. What is remarkable is that the fantastic, muddled churning up of society by the barbarians in the last fifty years has condemned this cultivated East European to spend his life driving a taxi in the plains of Galilee.

I had another similar example of the fantastic kaleidoscope of contemporary history in the mountains of Galilee. Twenty years or more ago there lived in Rodmell a man whom I will call Mr X. He was in a shipping company, a member of the Rodmell Labour Party, and a Jew, though at the time I did not know that he was a Jew. He was a gentle, intelligent man, very much interested in and keen to discuss contemporary problems. One day he came to me and told me that he was going to Israel to become an Israeli and a member of a kibbutz. Mr X disappeared from the Rodmell Labour Party, the Sussex downs which he loved, and the London shipping office which I doubt whether he loved at all. He became Avraham ben Yosef, the shepherd of a kibbutz on the mountains of Galilee looking into Syria, and every Christmas for years he wrote me a long and very interesting letter describing his life and experiences. During our tour of Israel we went up into the Galilean mountains and spent two or three days in Safed. There I found that we were only sixteen miles from Avraham ben Yosef's kibbutz, so I hired a taxi to drive us out to see him. The kibbutz was perched on the top of a mountain with a magnificent, though stern and stark, view over the foothills to the Syrian plain. The kibbutz itself was rather stern and stark. There was no luxury, but, as everywhere in Israel, a hum and warmth of life and energy. My friend lived in a gaunt room, which was a kind of outhouse in which the only ornament and amenity was a cold-water tap. Every day he took his flock of sheep down to the plain on the Syrian border, and there, though officially there was a state of bitter enmity and perpetual war between Syria and Israel, the Syrian shepherds from across the border came and fraternized with the Israeli, and the sheep grazed and the shepherds ate their lunch together.

I asked Avraham ben Yosef to come and have dinner with us at the hotel in Safed. He arrived at 7.30 carrying a large suitcase full of books and pamphlets which he wanted me to look at after dinner. At half-past ten he got up and said that he must go. I asked him how he was going to get back to the kibbutz sixteen miles away. He was going to walk, though, if the police patrol car happened to pass him, they would give him a lift. It was characteristic of the way of life in Israel today that a man who had been a sedentary worker in a London shipping office should think nothing of walking thirty-two miles, carrying a heavy suitcase on mountainous roads in order to meet someone at dinner.

Three things remain in my memory and I can still see them clearly when I think of Galilee. The first is Avraham ben Yosef setting off with his suitcase to walk sixteen miles over the mountain to his austere kibbutz. The second is the great silent wadi below Safed, so packed with wild flowers that we found twenty or thirty different species in the space of a few yards. The third is a regiment of crabs climbing up the staircase of the large hotel on the sea of Galilee. When we got to the hotel, they told us that they could only take us in for two nights, as after that they closed down altogether during the hot season. But the day after we arrived the Syrians opened fire on an Israeli fishing-boat and killed or injured one of the fishermen. This was an international "incident", and three officers of the United Nations Commission came up to investigate it. The hotel had to remain open for an extra day and we decided to stay on. No one else did and we and the three Commissioners were the only guests in this large silent hotel. We were sitting in the vast lounge hall after dinner when we saw suddenly a long procession of large and small crabs file past us and begin climbing up the staircase. When the hotel manager passed by us paying no attention to the long line of crabs, I asked him what it meant. He said that whenever the hotel closed down, it was immediately invaded by hundreds of crabs from the lake and they remained there, upstairs and downstairs, until the hotel reopened with the cool season.

I must mention one other unexpected conversation which I remember to have had when travelling. It was in Ceylon; the plane in which I was to fly back to England was held up, I think in Burma, for some repairs and we hung about all day long awaiting its arrival. At last we were told that it would arrive at midnight, and we drove out to the airport; but when we got there we found that there would still be a long delay. We sat outside the building in the gentle, drowsy, soft air of a tropical night. One of the Sinhalese employees of the airport came and sat down by me and began to talk, first about Ceylon and the years which I had spent in it as a civil servant and then, for some extraordinary reason, about philosophy. There are, I think, four sciences, departments of knowledge, or disciplines, as they are called in academic circles, which are almost completely phoney: the most bogus is theology; then comes economics; in third and fourth place I should put sociology and metaphysics, with little to choose between them. I was a little uneasy when I found that the airport official sitting by my side had a passion for the metaphysics of Kant and insisted upon my discussing with him what I think is called *Prolegomena to any Future Metaphysic*. For the next half-hour we forgot the non-existent plane and all other terrestrial things and lost ourselves, not in an *O altitudo*, but in the semantic fantasies and obscurities of the *Critique of Pure Reason*. I was soon metaphysically out of my depth, but the enthusiasm of my companion was such that it prevented him seeing my ignorance, and I had not the courage or the heart to reveal my misprision both of pure reason and any future metaphysic. At any rate, just as the Sea of Galilee will always remain inseparably connected in my memory

Some pleasures of old age: gardening; cats; and dogs

with crabs, so Colombo is now inseparably linked in my mind with Kant.

My journey to Ceylon was in February 1960; I went there with Trekkie for three weeks. It was half a century since I had spent seven years there as a civil servant, a period of my life which I have described in the second volume of my autobiography, *Growing*. I wanted to revisit, before I was too old or too dead to do so, the strange places where I had, rather absent-mindedly, in my youth helped to govern the British Empire. I set out with some misgivings, for two reasons. First, there had just been throughout the island some very serious riots in which the Sinhalese attacked the Tamils with considerable loss of life and destruction of property. I thought that after that kind of bloody disturbance there might be restrictions on travel in some areas, and I did not want to go to Ceylon unless I was free to wander about anywhere and everywhere. However, when I went to the High Commissioner's Office in London, they assured me that there would be no difficulty of any sort, and, in fact, not only was this so, but it was surprising to find very few traces of this horrible inter-racial conflict.

My other misgiving arose from the fact that imperialism and colonialism are today very dirty words, particularly east of Suez. I hoped, if I revisited Ceylon, to be able to go to the places where I had worked as a Government servant and see something of how, now that Ceylon was a sovereign independent state, their administration compared with ours. But to do this I would need to have some help from the Sinhalese and Tamil administrators of today, and I feared that I might find them, not unnaturally, contemptuous if not hostile. Would they not say, or at any rate think: "Fifty years ago you were here ruling us, an insolent, bloody-minded racialist and imperialist. Thank God we have now got rid of you and really we don't want to be reminded of how you lorded it over us and exploited us in the bad old days."

My fears were entirely unnecessary; I have never had such an enjoyable or interesting journey as my three weeks travelling up and down Ceylon. It was mainly due to the welcome and the help which I got from every Government officer from the Governor down to the eighty-three-year-old Aron Singho, who had been my peon in the Hambantota kachcheri fifty years ago. Their attitude was the exact opposite of what I had feared it might be. I visited the four places in which I had worked as a civil servant: Jaffna, Mannar, Kandy, and Hambantota. In each of these places the Government Agents took me in hand and went out of the way to show me exactly how they were administering the districts and provinces. Every Government Agent whom I met, except one, went out of his way to impress upon me the fact that things were better in our time than they are today. This was the kind of thing which they said to me everywhere: "In your time when you administered a district or a province things were really much better than they are for us today. You had no local axes to grind, nor had the central government in Colombo. When you were appointed Assistant Government Agent of the Hambantota District fifty years ago, all you were concerned with was the prosperity of the District—that is what you aimed at and worked for. And you were allowed to get on with the job—no one deliberately obstructed you; practically no one interfered with you; the central government in Colombo encouraged you. But for us today it is entirely different. For instance, here am I Government Agent administering the Province; I am trying to do what you British civil servants did: all I am after is the prosperity of my Province. But I am never left in peace to get on with my job, as you were; I am always being interfered with by the politicians. All I care for is the good of the Province, the good of the people here. All they care for is a vote. They will interfere and stop me doing something which will benefit the whole Province in order to win the votes of a few people who have some vested personal interest here. And if we oppose this kind of thing, the demagogues in Colombo try to put the people against us by saying that we civil servants are part of the old 'feudal system' of the wicked imperialist days. They have already abolished the old headman system as 'feudal', though if the truth be known they have abolished it in name rather than in fact: if you go into a village today, you will find someone doing exactly the same work as the village headman did fifty years ago, but he will be called by some newfangled title instead of vidane or village headman."

The Sinhalese seem to be naturally courteous and honey-tongued, and in the East people are far more likely to say nice things to you, if you are a stranger, than they would be in, say, France or Germany. But I do not think this tribute to administration under the British Empire was simply flattery. There was much to be said against the imperialism of the British Empire in the years from 1904 to 1911, when I helped to rule it light-heartedly in the Hambantota District of Ceylon, but there were also some very good things in it. All governments are evil in one way or another, and the worst thing about our rule in

India and Ceylon was its democratic hypocrisy, its failure to fulfil its democratic professions and to associate the people of the country with the government of the country; this applied particularly to the upper regions of power, prestige, and government. Contrary to what we professed, we never did anything to prepare the way for self-government or responsible government. Our manners, officially and socially, were often deplorable and nearly always arrogant. But in 1900 the population of Ceylon was almost entirely agricultural, peasants and cultivators living in villages. In these villages the standards of living, education, and culture were low. Given these conditions, our provincial administration had some very good points; it was honest and, though patriarchal and paternal, i.e. no doubt 'feudal', the civil servant at the top, as ruler, was concerned solely with what he thought to be the good of the people and of the Province. There is evidence of this in the fact that in 1960 many years after we had left Ceylon to govern itself this local provincial administration was exactly the same as it had been fifty years before when I left the island in the heyday of imperialism.*

It was not only the Ceylon civil servants of today who handed these bouquets to the old imperialist civil service of my day. At the end of my three weeks' visit, on the day

* I must add that I really agree with the Ceylon critics who object to the system as being "feudal". The rule of subject or colonial peoples by imperial powers through their own headmen used to be called the system of "indirect rule". In certain stages of economic and social development it was inevitable. But it was in many ways a very bad system and everything ought to have been done by us and by our successors by economic development and education to make it unnecessary.

when I left for England, the *Ceylon Daily News* had an
article which began as follows:

> Mr Leonard Woolf's presence here after a lapse of
> fifty years inevitably takes one's mind back to the public
> service of the colonial era. Immediately one remembers
> such names as Emerson Tennent, H. W. Codrington,
> Rhys Davies, Sir Paul Pieris, Senerat Paranavitane,
> and others who, like Woolf, not only worked conscien-
> tiously at their day-to-day tasks, but found time through
> their "extra curricular" research to make an essential
> contribution in such fields as history, literature, and
> oriental studies.
>
> One quality characterised the public service in
> this period—the ideal of service to the community.
> The public servants of this era were not afraid to
> move among and with the people. In this they
> provided a striking contrast to the successors of
> today. The latter have deliberately built a wall be-
> tween themselves and the people they are meant to
> serve.

I have a nostalgic and, I suppose, sentimental love of
Ceylon and its people. Ceylon and youth! Youth and the
sun and sand and palmyra palms of Jaffna; youth and the
lovely friendly Kandyan villages and villagers up in the
mountains; youth and the vast lone and level plain of the
low country in Hambantota, the unending jungle which
tempered in me the love of silence and loneliness. Youth
and the jungle! Does not my invocation show that they
hold in my heart and memory the place that youth and
the sea held in the memory of Conrad and his sea captains

and chief officers? Listen to the voice of Marlow and Conrad:

> Ah! The good old time—the good old time. Youth and the sea. Glamour and the sea! The good strong sea, the salt, bitter sea, that could whisper to you and roar at you and knock your breath out of you. By all that's wonderful it is the sea, I believe, the sea itself—or is it youth alone? Who can tell? But you here—you all had something out of life: money, love—whatever one gets on shore—and tell me, wasn't that the best time, that time when we were young at sea; young and had nothing, on the sea that gives nothing, except hard knocks—and sometimes a chance to feel your strength —that only—what you all regret.

Substitute "youth and Ceylon and the jungle" for "youth and the sea" and, no doubt, it might be the nostalgic voice and purple patch, the slightly lachrymose memory of Leonard Woolf instead of Conrad and Marlow—though in the bottom of my heart I do not really feel quite so nostalgic and lachrymose for "the good old time" as these sentimental seamen. But I did genuinely and profoundly feel something of all this for Ceylon and my youth, and to feel it again, not only in memory but in the sounds and scents of Kandyan villages and low country jungle and in the voices of Sinhalese and Tamils, was what made this three weeks' resurrection of time past so enjoyable.

It was also naturally extraordinarily pleasant to be received everywhere with a friendly and even affectionate welcome, and, though normally I much dislike official receptions and the pomp and circumstance of the upper crust in every society—from Kings and Dukes to the

Dictators of the Proletariat—I must admit to the discreditable enjoyment of being treated as a V.I.P. all over Ceylon. This was partly due to the desire of so many people to know what a person like myself who had had an intimate knowledge of the country half a century ago thought of the country today now that the people were governing it for themselves. Over and over again I was asked: "What do you think of Ceylon today? Has it changed much from your time?" But it is not just vanity if I say that my welcome was also due to other things; chief among them was *The Village in the Jungle*, the novel about Ceylon which I wrote in 1913. It has been translated into Sinhalese and a Sinhalese company is making a film of it. The book is still read in the island and has won me the reputation among many Sinhalese and Tamils of not only loving the country and sympathizing with the people, but also of understanding them. That reputation was enhanced by what happened in 1916 when for a year or more I worked closely with the delegates of the Sinhalese who came to London to try to get justice with regard to the riots of 1915.*

Another thing which told in my favour, rather surprisingly, was the official diary which I kept from August 28, 1908, to May 15, 1911, when I was Assistant Government Agent of the Hambantota District in Ceylon. All Government Agents and Assistant Government Agents of those bad old days had to keep a detailed day-to-day diary of what they did and send it each month to the Secretariat in Colombo, where it was read with some care. (Anything of particular interest—or discredit to the

* The details of this story, discreditable to the British Government, are given in *Beginning Again*, pp. 229-31.

diarist—would be shown to the Colonial Secretary or even to the Governor, and I was once severely rapped over the knuckles by His Excellency for including in my diary, with the inexcusable arrogance and sublime courage of youth, some sarcastic—and not unjustified—criticism of my superior, the Government Agent of the Southern Province.) These official diaries go back in some provinces for over a century and a half; when I was stationed in the Northern Province I found that the first diary in the Jaffna kachcheri was written by the officer who at the end of the eighteenth century victoriously entered Jaffna after Britain had seized Ceylon from the Dutch in 1795. For the historian of imperialism these diaries give an extraordinary day-to-day history of British administration in a Crown Colony for more than one hundred years. In my case I took a great interest in writing my diary fully and frankly. That is why I find it rather surprising that I gained kudos from it from Sinhalese and Tamils in 1960; for it was not written for publication or for the eyes of the public; it was a highly confidential day-to-day report to the central government. It was the kind of document which should show up the iniquities of the ancient imperialists' regime.

I think that in fact it does, for it gives an accurate and vivid picture of British paternal colonial government half a century ago; to the unprejudiced eye both the dark and the light spots in the picture are visible. Luckily for me in 1960, at any rate during my visit, people generously ignored the dark and gave me credit for the light. As soon as I arrived in Colombo, Mr Shelton Fernando, who was then head of the Civil Service, came to see me and presented me with a copy of my diaries from the Government.

Mr Fernando was, and still is in his retirement, a remarkable man. Many hard things are said about government servants—some of them no doubt justified—but all tip-top civil servants of the British species whom I have met have been admirable people, not only masters of their own strange and difficult art of administration and with the *robur et aes triplex circum pectus* without which in their position they could hardly maintain sanity or even life, but also men of exceptional humanity and civilization. Shelton Fernando, who was Permanent Secretary of the Ministry of Home Affairs—the equivalent of Permanent Secretary of the Treasury in Whitehall—was as good as any of them when I first met him in Colombo. It was the beginning of a very pleasant and instructive friendship, for he is a man of wide interests and has a passionate interest in everything connected with Ceylon and its history; for the last eight years I have received from him almost monthly a long letter about what is happening or has happened in the island.

Mr Fernando took us to see Sir Oliver Goonetilleke, the Governor, and Mr Dahanayake, the Prime Minister. The Governor said that, when we went to Kandy, we were to stay as his guests in his official residence, The Pavilion. We did so, and it was a curious experience. In 1907 I had been sent to Kandy as Office Assistant to the Government Agent, Central Province, and I remained there for a year.* I lived then in the O.A.'s humble bungalow almost at the gates of The Pavilion. Now we were in the rich man's palace, looking down on the poor man's cottage which I had previously lived in. The palaces which our imperialist Proconsuls inhabited in Asia were often remarkable

* See *Growing*, pp. 132-71.

buildings, and the Kandy Pavilion was in its way, apart from its imperial pomp and circumstance, rather beautiful. It is embedded in a lovely semi-tropical garden, whose loveliness, in my eyes, is enhanced by a large troop of gubernatorial monkeys who swing from bough to bough of the great trees, and, like Luriana Lurilee, "laugh and chatter in the flowers".

When I went to see the Prime Minister in Colombo he talked to me about my Hambantota diaries and later instructed Mr Fernando to have them printed and published by the Ceylon Government. They were published in Ceylon in 1961 through the *Ceylon Historical Journal*, and The Hogarth Press published them in England. When I read them in the Galle Face Hotel in Colombo fifty years after I had written them, they called up before my mind and almost my eyes the picture of what Ceylon had been (and what I had been) in those far-off days. I was very glad to have the opportunity of recalling that picture before setting off on my sentimental journey to Hambantota, Kandy, and Jaffna. It was strange to see how changed and in some respects how changeless the people, the government, the land, and even the jungle had become since I had last seen it. I gave an account of what seemed to me the most important changes in the preface which I wrote for the edition of the diaries, and instead of trying to do the same thing all over again, I will quote what I wrote there.

To give an account here of what those changes have been would mean writing, not an introduction but a book; but I can summarily just mention three immense changes which I observed. The first is the revolution which comes to a people when they win "independence", when they govern themselves. When I left the Civil Service in 1911, every Civil Servant administering in the provinces and districts was a European, indeed practically every member of the Civil Service was a European and so were nearly all the senior officers in all departments of government. Today the government, the administration, the public services are Sinhalese and Tamil from the highest to lowest. In other words the people of Ceylon today are governing themselves, instead of being ruled by young men (and old men) born in London or Edinburgh. The tempo of government and of life is quite different, it is more lively and vigorous in 1960 than it was in 1911, and that is mainly due to self-government.

The second change is also a change in tempo. When I was in the Civil Service, the motor car had hardly reached Ceylon. We travelled about our districts on a horse, on a bicycle, or on our feet. The pulse of ordinary life was determined by the pace of a bullock cart. There were no motor buses—even the "coach" from Anuradhapura to the Northern Province was a bullock cart. Today the pulse of ordinary life beats to the rhythm of the motor car or motor bus, 30, 40 or 50 miles from village to village and from town to town. This has of course some great advantages, but also, I think, some disadvantages. One result has been that the kind of jungle village described by me in *The Village in the Jungle* is ceasing, or perhaps has already ceased, to exist. At any rate it would be quite impossible for a Government Agent to become intimately acquainted with it as I did fifty years ago. You could only get to know the villagers and their villages by continually walking among them, sitting under a tree or on the bund of a tank and listening to their complaints and problems. Today one drives through the village at 30 miles an hour.

The third change is economic. In a visit of three weeks one cannot of course really learn what the economic conditions of a country like Ceylon are. But my impression everywhere was that the standard of life is on the average higher today than it was in 1911 and that "prosperity" is a good deal wider spread. At any rate the changes are great in a district like Hambantota. The poverty-stricken villages in the jungle, the Beddegamas of my time, have almost ceased to exist; where there were thousands of acres of waste land and scrub jungle, there are today thousands of acres of irrigated paddy fields, good roads, and flourishing villages. These changes are very great and all to the good. And yet beneath the surface there is much, I feel, that has hardly changed at all. I revisited some of the out-of-the-way villages which I had known so well, both in the Kandyan hills and in the low country. Gradually the people, adults and children, gathered round, stared at me, and began to talk desultorily, sometimes about the old days. I may be wrong, but it seemed to me that something of the old village typically Sinhalese life still goes on beneath the modern surface. There were many bad things in those old days, but there were also some good things. At any rate it is to the Sinhalese way of life and the Sinhalese people who lived it that I look back with a kind of nostalgic, and no doubt sentimental, affection.

I have said that I had an almost universal flattering welcome in Ceylon. I did have one extremely hostile reception, which shows that some human beings, like the elephant, never forget—never forget and never forgive. The day before my departure, I was sitting in the Galle Face Hotel talking to Mr Fernando when I was told that someone wished to see me. My visitor turned out to be Mr E. R. Wijesinghe, aged eighty-six; he had been a Mudaliyar or Headman of East Giruwa Pattu in the Hambantota District when I was Assistant Government Agent there fifty years ago. I had only a vague, misty remembrance of him. He now stood in front of me and Mr Fernando and recalled to my memory an incident of which again I had a dim and misty recollection. It had happened during a terrible outbreak of rinderpest which ravished the district. The Government had instructed me to see that all cattle were kept in enclosures or tethered and all infected beasts immediately destroyed, and I had handed on these instructions to all my headmen, including the Mudaliyar. One day I received information that there was a buffalo, badly infected with rinderpest, wandering about in a village which was within the Mudaliyar's area. I sent a message to him to meet me there next day in the morning with the village headman. The village was some twenty miles from my bungalow and I rode out there in the early morning. The Mudaliyar and the village headman met me and took me to the bund of the village tank which was quite dry. Some distance away across the tank was a buffalo which, they admitted, was badly infected with rinderpest.

I had brought a rifle with me and I told the village headman to go across the tank and drive the buffalo down

205

to me, so that I could shoot him. The headman went off but soon came running back, saying that the buffalo was savage, owing to the sores all over its head (the result of the disease) and would charge him. I gave my rifle to the Mudaliyar and told him that I would go and drive the buffalo down to him so that he could shoot it. This I did and the Mudaliyar shot the unfortunate beast, which was in a terrible state owing to the disease.

Under the bylaws made for dealing with the outbreak, it was an offence, punishable with a fine or imprisonment, to keep cattle untethered and also an offence not to destroy an infected beast, and the owner could be tried by the Police Magistrate. As Police Magistrate, I told the Mudaliyar that I proposed to charge and try the owner of this buffalo at once—"Who was he?" To my amazement I was told that the owner was the village headman standing hangdoggedly before me. I tried the owner of the buffalo for two offences—not tethering his cattle and not destroying an infected buffalo—and I fined him ten rupees. As Assistant Government Agent, I then tried the village headman for not carrying out his duties, i.e. reporting or prosecuting the offender (himself) for breaking the law, and I fined him ten rupees.

All this, which I had forgotten, the old Mudaliyar, standing in front of Mr Fernando and me in the Galle Face Hotel fifty years afterwards, recounted in great detail and considerable bitterness, and, as he spoke, looking back over the long procession of years, I suddenly remembered and saw vividly again the scene of us three standing in the sweltering heat in the parched and waterless tank with the dead buffalo swarming with flies, near the village with the magnificent name of Angunakolape-

Monks House: garden in evening

lessa. And when the Mudaliyar had finished his story, he fixed on me a beady and a baleful eye and said: "Was it just, Sir? Was it just? The village headman paid the ten rupees which you had fined him as Police Magistrate, but he could not pay the ten rupees which you had fined him for not carrying out his duties as headman. I had to pay it for him—I had to pay it for him. Was it just, I say—was it just, I ask you, Sir?" "Yes," I replied, "it was just. He had committed two entirely different offences, one as the owner of the buffalo and one as village headman, and I punished him for the one as Police Magistrate and for the other as Assistant Government Agent. Yes, Mudaliyar, it was just." Of course, it was just. I was quite certain that it was just that day, February 28, 1960, in the Galle Face Hotel with the indignant Mudaliyar facing me, and yet I was not entirely comfortable about it, and I am quite certain that fifty years before in 1910 when I stood in the village tank, faced by the Mudaliyar and the unhappy vidane, I had the same ambivalent feeling. This ambivalence with regard to law and order and justice in an imperialist society was one of the principal reasons for my resigning from the Civil Service. Where the ethics of government and public service are concerned, I am a rigidly strict puritan both for myself and for other people. One of the bases of civilization is honesty and justice in government and the officers of government; for a headman or any other government servant to commit an offence himself and conceal the fact that he had committed it, while prosecuting and punishing other people for the same offence, seems to me outrageous, and I thought and still think that it would have been unjust if I had not punished the headman for the two completely different offences committed by him, one as a private person and the other as a government servant. But there is a cliché—which normally irritates me because it is so often mouthed complacently by complacent judges—that a decision must not only be just, it must also be seen to be just. Even if the heavens had fallen upon our heads in the Angunakolapelessa tank or in the Galle Face Hotel, even if Jehovah or Gautama Buddha had appeared and proclaimed that it was just for me to fine the headman twenty rupees, neither the headman nor Mudaliyar E. R. Wijesinghe would have believed it. I find and found this profoundly depressing. Mr Wijesinghe had as good a right to his code of conduct as I had, and there was no real answer to his question: "Was it just, Sir, was it just?" I thought it was, but I was not prepared to spend my life doing justice to people who thought that my justice was injustice. I felt a certain sympathy for the vidane of Angunakolapelessa and even for Mr Wijesinghe who had to fork out his ten rupees. At any rate I resigned from the Ceylon Civil Service as long ago as 1911.

I said that the Mudaliyar was the only person who, during my revisiting Ceylon, showed me hostility or reminded me that I had been an imperialist. After my visit, however, there were several articles in the Ceylon papers attacking me for arrogant imperialist behaviour fifty years before. The untrue story that, as a young man in Jaffna, I had deliberately struck a Tamil lawyer in the face with my riding-whip was resurrected. On the other hand Mr Fernando and other people wrote defending me.

I must return for a moment, before ending this volume, appropriately to the subject of old age. Ever since Cicero wrote his *De Senectute* old men have written pompous platitudes about it, truisms which nearly always contain ten per cent of truth to ninety per cent of untruth. Or they have suffered from the complacent hypocrisy of Seneca which allowed him to write: "*Ante senectutem curavi ut bene viverem; in senectute, ut bene moriar*" ("Before I grew old I took care to live a good life, in old age my care is to die a good death"). A few years ago I had to make an after-dinner speech and I chose the subject of old age, and most of my friends who heard it complained that I talked as much nonsense about it as Cicero and Seneca. I do not think that they were right. Their chief complaint was that my picture of old age was absurdly optimistic, leaving out all the miseries and aches and pains of the old body, the old mind, and the withered soul. I can, of course, only speak from my own experience. There are lamentable things in one's physical and mental decay as one goes downhill from middle age to death. But there are compensations. I admit that physically I am unusually tough and have so far escaped the major miseries of the moribund body. On the other hand a positive pleasure comes from the fact that, in Britain, one enjoys great prestige merely from not dying. If only you grow old enough, you get immense respect, affection, even love, from English people. Queen Victoria and the great W. G. Grace are well-known examples of this senolatry. Victoria, who had been unpopular in middle age, became more and more beloved of her subjects the longer she lived as a selfish, bad-tempered old woman. Grace, it is true, was the greatest of cricketers, but his enormous fame and popularity was due less to his genius than to the fact that he could make a hundred not out when he was sixty not out. Anyone who lives to the age of eighty, by acquiring the merit of not being dead, enjoys this irrational approval and even affection. It is a valuable asset, because it provides the basis for good relationships with other people. It is one reason why it is easier for a man of eighty to understand and get on with the young than a man of fifty or sixty.

There are other assets of old age. The storms and stresses of life, the ambitions and competitions, are over. The futile and unnecessary and false responsibilities have fallen from one's shoulders and one's conscience. Even the false proverbs tend to become true for old people, for instance, that it is no good crying over spilt milk—after the age of eighty. One has learnt the lesson that sufficient for the day is the good thereof. And one can almost say:

> Grow old along with me!
> The best is yet to be,
> The last of life, for which the first was made.

And one can say again: "It is the journey, not the arrival, which matters."

INDEX

Aberconway, Lord, 76n
Abrahams, William, 104
Anders, Gen. Wladyslaw, 67
Angell, Norman, 156
Anglo-Soviet Society, 151
Anon, 75
Armenian massacres, 21-4, 26, 168
Arnold-Forster,Will,156,160-1
Ashcroft, Peggy, 82
Attlee, Clement, 152, 159, 165
Auden, W. H., 103-5
Austen, Jane, 134
Avraham ben Yosef, 190-1

Backsettown Trust, 84
Baldwin, Stanley, 163
Barbarians at the Gate, 11
Barker, A. L., 116
Beginning Again, 79, 117, 121, 149, 153, 160, 182, 199n
Bell, Clive, 177
Bell, Julian, 103
Bell, Vanessa, 42, 81, 90, 93
Besterman, Theodore, 27n
Between the Acts, 40, 44, 52, 74-5, 107
Bevin, Ernest, 138
Bodley Head Press, 104
Bonham-Carter, Lady Violet, 139
Bowen, Elizabeth, 87

Brailsford, H. N., 156
Brecht, Bertolt, 106
Bunin, Ivan, 98, 125
Buxton, Charles, R., 156, 161-2, 165

Calas case, 27, 168
Cameron, Mrs Julia M., 81
Campbell, Jock, 153
Cecil, (Viscount of Chelwood) Lord, 161
Ceylon, 124-5, 130, 133, 142, 148-9, 153, 158, 186, 192; 1960 visit, 193-208
Ceylon Historical Journal, 202
Chamberlain, Neville, 28, 163
Chamson, André, 104
Charity Organizations Society, Hoxton, 153
Chatto & Windus Ltd., 112-13, 123, 179, 181-2
Churchill, Sir Winston S., 163-5
Civil Service Arbitration Tribunal, 132-5, 149, 151
Codrington, H. W., 197
Cole, G. D. H., 136-8, 153
Cole, Margaret, 137
Cole, Mrs (Headmistress), 22-3
Colefax, Lady Sybil, 53
Communism, 12, 13, 138, 168
Conrad, Joseph, 197-8
Creech Jones, Arthur, 162

INDEX

INDEX

INDEX

Vietnam, 169

Village in the Jungle, The, 199, 203

Voltaire, 20, 27n, 168

Walpole, Hugh, 81

War (1939-45), 9-11, 28-30, 46-9, 52-3, 163; evacuees, 30-1; air battles over Sussex, 31-8; 1940 setbacks, 54-6, 58; bombing of England, 59-64; boredom of, 64-5

Warner, Rex, 104

Waves, The, 41

Webb, Sidney, 152, 159, 165

Weizmann, Chaim, 185

Whitley Council, 133

Wijesinghe, E. R., 205, 208

Wilberforce, Octavia, 80-7, 90-92

Wilson, Harold, 153

Wittgenstein, Ludwig J. J., 48

Woolf, Leonard, attitude to totalitarianism, 9-12, 16-17, 28; considers suicide in event of war, 15, 45-6; on cruelty, 18-25; on injustice, 26-8; joins fire service, 34; on Virginia's writing, 41-3; and Virginia's death, 44, 69, 72-73, 93-6, 127, 149, 177; on imbecile children, 49-52; on class struggle, 61n; and wartime officers' visits, 66-8; on digression in autobiographies, 68-9; and servants, 69-70, 99-100; on death, 73, 177; on male dominance, 75-7; lectures, 77-8; and Hogarth Press, 97-102, 108, 110, 122-125, 180-1; as editor of journals, 102, 131, 136-46; on small-scale publishers, 108-110; on John Lehmann, 114; his Jewishness, 127-9, 166; his hard work, 128-31, 148; on 'magnetic fields' of ego and institutions, 143-8; political and committee activities, 150-4, 157-8; assessment of own achievements, 158-61, 166, 168, 171-2; on war and international government, 154-6, 163-6; on imperialism, 161-2, 164, 169, 194-6; on civilized ideals, 166-8, 183; on pleasures, 182-3; on old age, 183, 208-10; on writing, 183; on travel, 184-5; Zionism and Israel, 185-90; *Ceylon Daily News* on, 197; Ceylon diaries of, 199-202

Woolf, Virginia, and war, 15, 30, 32-4, 45-6, 49, 54; approach to writing, 40-4, 52, 74, 90, 175; sensitivity to criticism, 41; depression and suicide, 44, 69, 72-4, 77-80, 86-96, 127, 149; visits Penshurst, 56, 58; as book collector, 65-6; and

216

Books by Leonard Woolf
available in paperback editions
from Harcourt Brace Jovanovich, Inc.

SOWING: AN AUTOBIOGRAPHY OF THE YEARS 1880 TO 1904

(HB 319)

GROWING: AN AUTOBIOGRAPHY OF THE YEARS 1904 TO 1911

(HB 320)

BEGINNING AGAIN: AN AUTOBIOGRAPHY OF THE YEARS 1911 TO 1918

(HB 321)

DOWNHILL ALL THE WAY: AN AUTO-BIOGRAPHY OF THE YEARS 1919 TO 1939

(HB 322)

THE JOURNEY NOT THE ARRIVAL MATTERS: AN AUTOBIOGRAPHY OF THE YEARS 1939 TO 1969

(HB 323)

FIVE-VOLUME BOXED SET

(HB 324)